398

JUN ~ ~ ~ ~~~~
2/08

CALIFORNIA NATURAL HISTORY GUIDES

DRAGONFLIES AND DAMSELFLIES OF CALIFORNIA

California Natural History Guides

Phyllis M. Faber and Bruce M. Pavlik, General Editors

DRAGONFLIES
and DAMSELFLIES
of California

Tim Manolis

UNIVERSITY OF CALIFORNIA PRESS

Berkeley   Los Angeles   London

For Anne, Annette, and Ellen

California Natural History Guides No. 72

University of California Press
Berkeley and Los Angeles, California

University of California Press, Ltd.
London, England

Library of Congress Cataloging-in-Publication Data

Manolis, Tim, 1951–.
    Dragonflies and damselflies of California : Tim Manolis.
        p.   cm.—(California natural history guides ; v. 72)
    Includes bibliographical references (p.).
    ISBN 0-520-23566-5 (hc. : alk. paper).—ISBN 0-520-23567-3 (pbk. : alk.
paper)
    1. Dragonflies—California.   2. Damselflies—California.   I. Title.   II.
California natural history guides ; 72.

QL520.2.U6 M36   2003
595.7′33′09794—dc21   2002032189

Manufactured in China
12   11   10   09   08   07   06   05   04   03
10   9   8   7   6   5   4   3   2   1

The publisher gratefully acknowledges the generous
contributions to this book provided by

the Moore Family Foundation
Richard & Rhoda Goldman Fund
and
the General Endowment Fund of the
University of California Press Associates.

CONTENTS

Plate section follows page 122

PREFACE

The need for a guide like this was immediately apparent when I first became interested in dragonflies. Although I quickly found the excellent papers of Clarence Kennedy (1915, 1917) in a local library, these were considerably out of date, as was the most recent complete treatment of the California fauna by Pritchard and Smith (1956) (see the "References" section for this and other publications mentioned here). Even those sources that covered the odonates of North America were out-dated.

Interest in dragonflies has since blossomed, and a number of useful references have emerged in recent years. The Internet has also become a major forum for information and the exchange of ideas about dragonflies. I have not listed specific Web sites in this book because they come and go, or change addresses, with some frequency. Suffice it to say, a search on "California dragonflies" using any of the major search engines will get you well on your way. Despite this increase in publicly available information, the need still remains for a comprehensive field guide to California's dragonflies. Through the good graces of Kathy Biggs and Doris Kretschmer, I was given the opportunity to produce this one.

Considerable time was spent doing field work and visiting institutional and personal collections throughout the state for the groundwork for this book. Many people helped in this.

Kathy and David Biggs, Jeffrey Cole, Bruce Deuel, Rosser Garrison, Annette Manolis, Kathy Nakashima, Joseph Ramirez, Andrew Rehn, and Bruce Webb provided direct assistance and enjoyable company in the field. For access to collections and information about specimens or photographs in collections I could not personally visit, I thank Cheryl Barr, Robert Behrstock, Jutta C. Berger, Kathy Biggs, Michael Camann, Jeffery Cole, Thomas Donnelly, Sidney Dunkle, Rosser Garrison, Gregory Grether, Eric Hochberg,

Mark Holmgren, Alvaro Jaramillo, Lynn Kimsey, David Kistner, Michael Klein, Ron LeValley, David Nunnallee, Michael Patten, Dennis Paulson, Norman Penny, David Pryce, Brian Quelvog, Nathan Rank, Andrew Rehn, Kathy Schick, William Shepard, Janice Simpkin, Hal White, David Wyatt, and Douglas Yanega. Sarah Blanchette, Roger Jones, Dianna McDonell, Barbara Tatman, John Trochet, and Stan Wright provided access to areas that would otherwise not have been open to me. Dennis Paulson and Rosser Garrison, in particular, showed great patience in answering my many questions and graciously sharing their unparalleled knowledge of West Coast Odonata.

Production of this book would not have been possible without thoughtful reviews of early drafts by Robert Cannings, Rosser Garrison, and Dennis Paulson, or the staff at the University of California Press, especially Barbara Jellow, Doris Kretschmer, and Scott Norton. Work on this project has greatly benefited from the support of family, especially Annette Manolis, my soul mate, as well as Ellen Manolis, Ruth Price, Bill and Cassandra Walters, and Georgia and Ted Econome. I apologize to anyone who has helped me whom I forgot to mention.

INTRODUCTION

So, you are interested in dragonflies? There are certainly many reasons you might be. There is beauty in their bright colors and intricate wing veins. Their shape is unmistakable and correlates with equally unique and interesting behavior. Perhaps most fascinating is that they play out the adult stages of their life cycle in fine weather and in plain sight, where easily watched. This book introduces you to California's dragonflies—where they live, how they can be identified, and something of their habits.

The common name *dragonfly* can be applied to all insects in the order Odonata. They may also be called odonates. This name is derived from the Greek word for tooth and apparently refers to dragonflies' impressive chewing mouthparts. The Odonata order is a very old and distinctive group. Modern-type odonates are known from the Jurassic era (nearly 200 million years ago), and similar forms go back as far as the Carboniferous period (250 to 300 million years ago). Permian fossils have been found of giant dragonflies (Protodonata), among them the largest known insects, with wing spans of over 2 ft. Although considerably smaller, modern dragonflies remain among the dominant aerial predators in the insect world.

All adult odonates share a suite of characteristics that are distinctive and render them instantly recognizable:

1. Two pairs of relatively large wings of about equal length, with an extensive and intricate network of veins
2. Large compound eyes
3. Minute and inconspicuous antennae
4. A relatively long, often narrow, 10-segmented abdomen with, in males, terminal appendages used for grasping females when mating and secondary genitalia for sperm transfer under the second segment (this latter feature is unique to the group)

This basic design has produced such a finely tuned aerial predator, with superb vision and unmatched agility on the wing, that it has undergone virtually no significant modifications for millions of years.

Within the order Odonata are three suborders: Anisoptera, Zygoptera, and Anisozygoptera. Anisozygopterans comprise a small group of two species, found only in Asia, that show a mix of characteristics found in the two other suborders.

Zygopterans are also commonly known as damselflies. Forty

species of damselflies occur in California. They have a characteristic body shape: a long, slender abdomen and a "hammer head," with large compound eyes widely separated on short stalks on either side of the relatively small face. When perching, most of our species carry their wings flat together, or nearly so, above the abdomen. Exceptions to this rule are members of the spreadwing family (Lestidae), which perch with wings held out to the sides, somewhat like anisopterans. The wings of all but two of our species (in the broad-winged damsel family [Calopterygidae]) are distinctly stalked (petiolate) at their bases, and the fore and hind wings are of similar size and shape.

Anisopterans are what most people consider typical dragonflies. Sixty-eight species are known in the state. They are more robust than damselflies and perch with their wings spread. The hind wings are broader at their bases than are the fore wings, prominently so in some species. The large compound eyes are closer to each other than they are on damselflies and even come in contact atop the head in many species. The face is relatively large and flattened.

Adult Dragonfly Anatomy

Odonates have the basic insect body plan, which consists of a hard exoskeleton fashioned into a head, a thorax, an abdomen, six legs, and four wings. Within this structure, certain details of anatomy, some unique to the group, must be learned in order to readily distinguish among similar species and to properly appreciate observed behavior. I have tried to simplify the descriptions and limit unfamiliar terms as much as possible. See the glossary, figs. 1 and 2, and the following descriptions for help.

Head

The head is relatively large and dominated by two large compound eyes that are oriented dorsolaterally (upward and to the sides). Each compound eye is composed of thousands of individual lenses, or facets. The color of these eyes in life may be quite striking and is typically darker above than below, often with a sharp break between the two regions. Black spots or lines on the eyes, some-

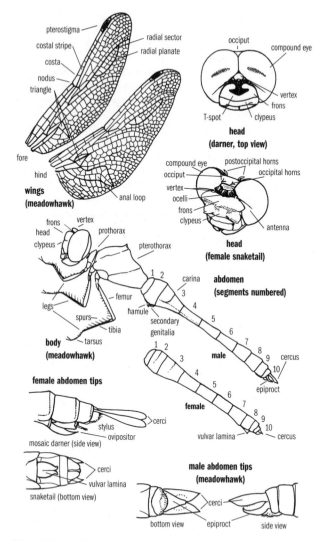

Figure 1. Dragonfly anatomy.

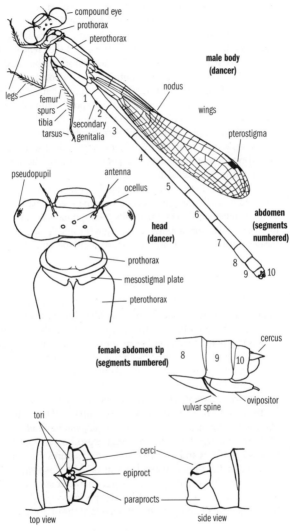

Figure 2. Damselfly anatomy.

times referred to as pseudopupils, are areas of enlarged facets pointed in a particular direction, which is the direction of most acute vision. The front of the face has chewing mouthparts below and is bordered above by a central plate called the clypeus. This plate is separated from the forehead region, a frontal lobe called the frons, by a horizontal groove (suture line). In some dragonfly species, the frons has a distinctive color pattern that resembles a bull's-eye or a capital T (referred to as a T-spot). Behind the frons and between the front of the compound eyes is the vertex, a plate bearing three simple eyes (ocelli). The relatively inconspicuous antennae arise from near the front of the vertex close to the compound eyes. A central plate at the upper rear margin of the head between the eyes is called the occiput. On dragonflies that have compound eyes in broad contact atop the head, the occiput is reduced to a small, triangular region. In some species, particularly some members of the clubtail family (Gomphidae), horns on the top (occipital horns) and the rear margin (postoccipital horns) of the occiput are useful identification features in the hand.

Thorax

The thorax can conveniently be divided into the prothorax, a small segment attached to the head and bearing the front pair of legs, and the pterothorax, the large, sturdy midsection that houses the flight muscles and bears the wings and remaining two pairs of legs. The pterothorax—often simply referred to as the thorax in this book—is strongly slanted (upper surface backward, under surface forward) so as to project the wings backward and the legs forward. This slant means that what is structurally the front of the pterothorax appears to be on top of the structure. Damselflies have a pair of mesostigmal plates at the leading edge of the front of the thorax just behind the prothorax. On females, these small plates and the prothorax are structures that interact with the male's caudal appendages in tandem linkage, so they often feature important species-specific characteristics useful for in-hand identification. Some odonate species have a distinctive tubercle, or small bump, on the underside of the pterothorax.

Legs

The legs of odonates consist of the typical, jointed arthropod segments, most notably the femur, tibia, and tarsus. Each tarsus consists of three segments and is tipped with two claws. Spiny spurs line the margins of the femur and tibia. Because of their forward positioning, the legs are of little use for walking; instead, they are used for perching, scooping up and handling prey, and (in the case of the front legs) for grooming the eyes and face.

Wings

The large, many-veined wings of dragonflies bear many distinctive features. The complex terminology developed for wing venation is, however, beyond the scope of this guide. Key features you should become familiar with, illustrated in fig. 1, are the anal loop, the triangle, the costa and costal stripe, the nodus, the radial planate and radial sector, and the pterostigma. The pterostigma, a thickened, often distinctively colored wing cell along the leading edge of the wing near the tip, apparently serves an aerodynamic function and, in some species, is also an element in visual displays or signaling.

Abdomen

The abdomen is divided into 10 segments, numbered from 1 at the base to 10 at the tip. The major features of the abdomen useful for identification are the sexual appendages, extending from the terminal segments, and the secondary genitalia of males. The latter are located under abdominal segment 2. Of the many intricate structures that make up the secondary genitalia, the most useful to learn for field identification are the hamules, which are hooklike structures on most dragonflies; they are visible with a hand lens and are often species specific in shape.

The terminal appendages of males, which are used to grasp females by the rear of the head (in typical dragonflies) or the thorax (in damselflies) during mating, are often important features for distinguishing look-alike species. It is therefore worthwhile to learn the names and positions of these structures.

The cerci (singular, cercus) are the pair of upper (superior or dorsal) appendages. When the male curls his abdomen down and forward to grasp the female, the cerci curl under the upper rim of the head (in typical dragonflies) or contact the mesostigmal plates (in damselflies). Male cerci are often relatively large appendages with hooks or spines for grasping. The cerci of females are typically simple, cone-shaped or leaflike structures and only occasionally of use in field identification.

The lower (inferior or ventral) grasping appendages of male damselflies and typical dragonflies are different in nature. Male damselflies have a pair of structures, the paraprocts, that grasp the prothorax during tandem linkage. In some species, these are relatively large and strongly hooked. In nearly all other odonates, the paraprocts are relatively inconspicuous, small, rounded structures. The inferior grasping appendage of typical male dragonflies is a single epiproct (strongly forked in some species) that grasps the top of the head or eyes of females. This structure typically is not a conspicuous feature on female dragonflies or most damselflies. However, on male dancers *(Argia),* the shape and size of the epiproct—which appears as a small lobe project-ing rearward from the upper middle rim of segment 10—relative to the shape of the projecting pads (called tori [singular, torus]) on either side of it are of use in identifying some species.

The females of all damselflies and some dragonflies (darner and petaltail families [Aeshnidae and Petaluridae]) have a fully formed ovipositor, which is a complicated structure containing paired valves and cutting blades, on the underside of abdominal segments 8 and 9. The ovipositor is used to insert eggs into plant tissue, mud, or other substrate. Some species have a stylus, which is a thin, needlelike projection, at the end of each of the two valves of the ovipositor. Species without a true ovipositor (most of the typical dragonflies in our area) have a more or less well developed vulvar lamina, a plate that extends rearward from segment 8 to cover part of the undersurface of segment 9. This plate, which is often distinctly bilobed, may be used to carry egg masses or oth-erwise aid in the dispersal of eggs. In the spiketail family (Cord-ulegastridae), the vulvar lamina is highly modified to form a spikelike structure that inserts eggs, much as a true ovipositor might, into aquatic substrates. Some species of damselflies have a vulvar spine on the rear lower margin of segment 8 that projects over the genital opening at the base of segment 9.

You will occasionally find odonates, especially damselflies, with tiny red "beads" attached, often in small clusters, to the un-dersurface of the thorax or abdomen. These are not part of the odonate but rather are larvae of parasitic water mites, which hitch a ride on odonate larvae and then make the transfer to the adult form at the time of emergence. The mite larvae attach themselves to the hardening body, sucking fluids from their host. When the adult odonate later comes into contact with water—

for example, during oviposition—the mites detach and return to the water to complete their life cycle.

There are few other insects that might be mistaken for odonates. Perhaps most likely to cause confusion are adults of the antlion family (Myrmeleontidae, order Neuroptera), which resemble adult damselflies but have noticeably longer, club-tipped antennae.

Dragonfly Behavior

Adult dragonflies use vision as their primary means of assessing their environment. In this way, they are like us, and their behavior, as compared to that of many other, more secretive insects, is relatively easy to understand if we simply watch what they do. Many specific behaviors are characteristic of particular species or groups of species, so in making an identification, observing behavior is often as important as noting appearance.

Dragonfly behavior has evolved in response to a few simple needs:

The need to eat
The need to avoid being eaten
The need to reproduce
The need to regulate body temperature (thermoregulation)
The need to disperse

Some of the distinctive behaviors odonates have evolved to meet these needs are discussed in the following sections.

Feeding Behavior

Dragonflies are voracious predators; they eat just about any animal they can catch and chew, including other dragonflies. Most prey of adult dragonflies are flying insects, taken on the wing. The two general types of aerial feeding used by dragonflies are hawking (the constant pursuit of flying insects) and sallying (darting out from a perch to capture prey and then return to the perch). Hawking dragonflies remind bird-watchers of swifts or swallows and often feed in swarms as do those bird species, whereas salliers are reminiscent of flycatchers. Some species are hover-gleaners, picking prey from vegetation and other substrates while in flight. Most species use one of these foraging strategies predominantly but may occasionally use the other two as well.

Dragonflies that typically hawk for food include the darners, river cruisers *(Macromia)*, baskettails *(Tetragoneuria)*, emeralds *(Cordulia* and *Somatochlora)*, spiketails, gliders *(Pantala)*, and saddlebags *(Tramea)*, among others. They are strong fliers of medium to large size. In genera such as the gliders and saddlebags, the hind wings are relatively broad based, allowing for almost effortless, gliding flight in light winds. Hawking dragonflies frequently fly back and forth along a set path or series of paths over open fields and meadows or along creeks, rivers, and even roads. They are on the wing for extended periods of time. Darners, gliders, saddlebags, and others that hawk may form feeding swarms of dozens to hundreds of individuals, most often at dusk in late summer and fall. When they eventually perch, they do so to rest, digest a meal, or avoid unfavorable weather conditions. In general, they are somewhat cryptically colored (dull earth tones predominate), tend to perch high or in the shade of dense vegetation rather than on low, exposed perches, and typically hang from a perch, their bodies oriented vertically.

Many species in the skimmer subfamily (Libellulinae) and many damselflies are salliers. From a perch on the ground, vegetation, fence post, or other surface, they alertly scan the sky for potential prey. Typically they sit in exposed situations that provide a wide field of view. When they spy a meal, they dash out to capture it and return to a perch, usually the one they just left, to finish eating. When perched and actively foraging, they tend to adopt a flight-ready position, the body oriented horizontally.

The third feeding technique used, especially by damselflies, is hover-gleaning, which involves flitting from spot to spot, picking food items off vegetation in rapid bursts, followed by a brief period of perching to chew up the prey. American bluets *(Enallagma)* are often seen feeding in this way. Other damselflies, such as the spreadwings and broad-winged damsels, sally out from a perch to fly catch or hover-glean a single prey item at a time, then perch again to finish eating.

Antipredator Behavior

Dragonflies typically avoid aerial predators—birds, bats, and insects such as robber flies, wasps, and even other dragonflies—by agile aerial maneuvering, as anyone who tries to net them can attest. Disturbed damselflies frequently dodge into nearby vegetation. If flight is a less viable option (e.g., at cold temperatures), perched damselflies sidle around perches such as grass stems or small branches, using the perch as a screen much as woodpeckers use tree trunks.

The patterns on the bodies or wings of some species may serve as cryptic coloration against certain backgrounds, and this may in part influence perch selection. The bright colors of some species, especially the blues of darners and many damselflies, fade to gray at cooler temperatures, when mobility is reduced. Dragonflies also seem to magically disappear from conspicuous perches when the sun goes behind a cloud, perhaps to avoid detection by predators when their activity levels drop.

Reproductive Behavior

Reproduction is the major goal of an adult dragonfly's existence. After a few days or weeks of prereproductive life, during which it must feed, grow, and mature, it begins a programmed series of activities focused on reproduction.

The first step is finding a mate. In most cases, mates are sought near or at the body of water in which eggs are to be laid. In a few cases, mates are sought away from water and then escorted there. Males typically arrive at rendezvous sites before females. Peak mating hours vary among species but are often in the late morning or early afternoon. Males commonly interact aggressively with other males in order to establish territories or otherwise secure advantageous positioning for attracting or finding mates. Males of some species seek females from a perch, whereas others, such as darners, patrol in search of mates.

Courtship is rare in dragonflies, and males usually quickly pounce on females that arrive at rendezvous sites or are otherwise encountered during mate searches, even knocking them to the ground in some cases. They then quickly proceed to the next step, which is formation of tandem linkage.

Tandem linkage is the physical link of the male's abdominal appendages to the rear of the head or the thorax of the female. This linkage often provides a close fit of species-specific body parts, which may inhibit interspecific mating attempts (and coincidentally makes these body parts useful to humans attempting to identify individuals as to species).

Next is copulation, which also tends to follow quickly. Dragonflies are unique among insects in that the secondary genitalia of males are housed in the undersurface of the second abdominal segment. Usually after linking with a female, but sometimes before, a male transfers sperm from near the tip of the abdomen to a storage

Figure 3. A pair of Western Meadowhawks copulating in the wheel position.

area in his secondary genitalia. The female subsequently bends her abdomen forward to align her reproductive structures under the eighth abdominal segment with the male's secondary genitalia. This position, which involves two points of linkage, is called the wheel (fig. 3).

The elaborate complex of secondary genitalia in males not only stores and transfers sperm but is designed to remove any sperm placed by other males in prior mating attempts. Indeed, much of the time spent by a pair in the wheel (a few minutes to hours in some species) is taken up by sperm removal, followed by a relatively brief period of sperm transfer. Because the secondary genitalia also require a good fit, they, too, are useful for distinguishing a number of look-alike species.

After sperm transfer, the next step is oviposition. The female of some species (damselflies, darners, petaltails, and spiketails) uses her ovipositor, or vulvar lamina, to insert eggs into a substrate—usually some sort of vegetation. Other species use a variety of techniques, discussed in the species accounts, to drop or deposit their eggs in water, on vegetation, or on the ground.

The male may remain in tandem with the female during oviposition, apparently protecting his investment in the eggs being laid. In other species, the male hovers near the ovipositing female and chases off intruders (fig. 4), especially other males of the same species. The female of some species typically oviposits while alone. There is considerable variation within species in these modes of oviposition, however, and some species exhibit more than one type,

Figure 4. The male Common Whitetail hovers over an ovipositing female to guard her from other males.

depending on circumstances such as population density and habitat structure.

Thermoregulation

Dragonflies use a variety of behaviors in order to maintain an appropriate body temperature. Many of these movements and postures are designed to take advantage of solar radiation.

The most obvious of these is basking—perching on vegetation, fence wires, the ground, rocks, and other sites fully exposed to the sun, much as lizards do. The wings may be held down toward the sides to trap warm air close to the thorax. Many species that live in cooler climes, such as whitefaces *(Leucorrhinia)* and emeralds, have mostly blackish bodies, presumably to enhance absorption of solar radiation.

Some of the larger, hawking species, such as darners, can warm up by rapidly vibrating the large flight muscles in their thorax, either while perched or by flying.

Because dragonflies are most active in warm, sunny weather, they also have to worry about overheating. Simply seeking shade and reducing activity for a time can accomplish this. Some exposed perchers adopt a very distinctive position called the obelisk, in which the abdomen is pointed directly at the sun (nearly straight up at midday) to minimize the body surface area exposed to direct rays (fig. 5). The tip of the abdomen can also be pointed down (away from the sun) to achieve a similar effect; this posture is often adopted by saddlebags in flight, the dark patches on their hind wings shading the drooped abdomen.

As discussed in the section "Antipredator Behavior," the blue colors of many darners and damselflies and the red colors of some species such as

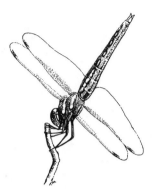

Figure 5. A Variegated Meadowhawk in the obelisk position.

meadowhawks *(Sympetrum)* are subject to reversible, temperature-induced changes, becoming bright at higher temperatures and dull at lower ones. The brighter color produced by higher ambient temperatures is also more reflective (absorbing less light), thus helping to reduce body temperature. Conversely, individuals at cooler temperatures increase their absorption of solar radiation via darker body color.

Dispersal

After emergence from the final larval stage, virtually all odonates disperse. For many, this involves flying a distance from a few feet to a few miles away. Such short-range dispersal probably serves a number of functions, including (1) occupation of good foraging areas, (2) avoidance of harassment by breeding adults, and (3) potential discovery of new breeding sites. Once they become sexually mature, adults return to breeding sites, from which they may commute between feeding and roosting sites.

A few species are capable of long-distance movements, although the exact nature and extent of these migrations is poorly known. Dragonflies believed to migrate in western North America are the Common Green Darner *(Anax junius)*, Variegated Meadowhawk *(Sympetrum corruptum)*, gliders, and saddlebags. Observations suggest that, in late winter and early spring, these species begin to emerge in large numbers in Mexico and the southern border states (including the warmer areas of California) and move north into the northern United States and southern Canada. They breed in summer and then die. A late summer and fall emergence resulting from this breeding activity typically produces large numbers of offspring that migrate back south to breed in fall and early winter, and their offspring in turn emerge in spring to repeat the cycle.

Life Cycles and Larvae of Dragonflies

Dragonflies are amphibians in the same general sense as frogs, toads, and many salamanders. The familiar, winged adults that are the primary focus of this guide are the final, reproductive stage in the odonate life cycle. But, like amphibians, they are preceded by an aquatic larval stage that, from hatching of the egg to emergence of the adult form, involves much of each dragonfly's total life span, growth, and development (fig. 6).

Although much less conspicuous than and markedly different in appearance from adults, odonate larvae (sometimes referred to as nymphs or naiads) are unique and fascinating in their way. Overall, larvae coloration is typically drab and designed to camouflage. Their eyes are smaller and their antennae are frequently more prominent than those of adults. Unlike adults, they use their legs for getting about, not for prey capture and handling. The abdomen is relatively short and sometimes armed with spines or knobs along the top and sides. In later larval stages, pads housing the developing wings lie on top of the front of the abdomen.

Like adults, larvae are high-level predators, feeding on a wide range of aquatic invertebrates, including other odonates. Large, active larvae are capable of capturing and subduing small fish and tadpoles. Their most distinctive feature, found only in the Odonata, is a double-hinged labium, or lower jaw. The labium consists of the postmentum, folded back under the front of the body; the prementum, hinged to the postmentum and, at rest, folded forward to cover it; and labial palps hinged to the front of the prementum and often covering the lower face. When potential prey draw within reach of this potent weapon, it is thrust forward at high speed. The movable palps at its tip, armed with hooks, teeth, and spiny hairs, capture and hold the prey. The labium is instantly retracted after capture, drawing the prey back to the chewing mouthparts. The structure of the labium varies among odonate families and is often useful for identifying larvae (fig. 7).

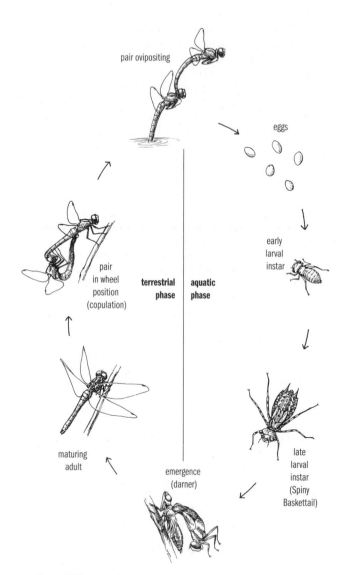

pair ovipositing

eggs

early
larval
instar

**terrestrial
phase** **aquatic
phase**

pair
in wheel
position
(copulation)

late
larval
instar
(Spiny
Baskettail)

maturing
adult

emergence
(darner)

Figure 6. Life cycle of a dragonfly.

Other unique features of odonate larvae are the gills they use to extract oxygen from the water in which they live. In damselflies, this is accomplished primarily by three leaflike gills that extend from the tip of the abdomen. Typical dragonflies have internal rectal gills over which they are constantly pumping water. Both types of gill systems also aid in locomotion, although in different ways. Some zygopterans can use their external gills as a sort of tail fin, swished from side to side to help them swim along. Anisopterans can rapidly expel water out the rectum to jet forward when a quick getaway is called for.

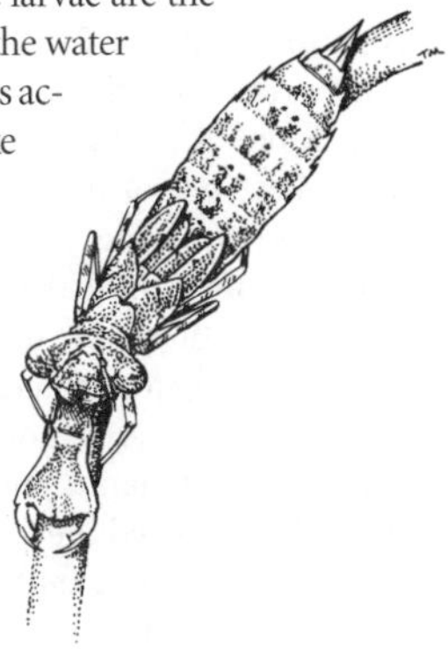

Figure 7. Variable Darner larva with labium extended for prey capture.

The basic larval body plan is modified in different species that live in different habitats and have different lifestyles. For example, dragonfly larvae that clamber about in aquatic vegetation and actively stalk prey, such as the larvae of large darners, have smooth, streamlined bodies and large eyes facing to the side. Species whose larvae sprawl in bottom sediments, such as the Pacific Spiketail *(Cordulegaster dorsalis),* have hairy bodies to which camouflaging detritus can adhere and eyes raised above the medium in which their bodies are mostly hidden. Shallow burrowers, such as the clubtails, have somewhat flattened, hairy bodies and thickened or platelike antennae that rest at the surface to detect prey. The elongate tip of the abdomen is also raised up above the surface of the mud to allow respiration through the rectal gills. Sprawlers and burrowers are ambush predators that sit and wait for prey to wander into range.

Most of California's odonates have a single generation per year. Adults emerge, mature, and lay eggs in the warmer months, primarily April through October. Eggs hatch within a few days or weeks, and the larvae grow through a series of about 10 to 15 molts. The stages between molts are called instars. Larvae usually overwinter in a relatively late stage of development. In the spreadwings and many meadowhawks, which typically breed in temporary habitats late in the season, it is the eggs that usually overwinter, hatching in spring. Adults emerge again the following year to repeat the cycle.

There are exceptions, however. Some smaller damselflies, such as forktails *(Ischnura)* and bluets, have long flight seasons and may have two or three broods per year, the last brood of the season overwintering as larvae. On the other hand, some dragonflies, often those that live in streams or rivers, at high elevations, or in other more demanding habitats, may live as larvae for 2 to 4 years before emerging. Some darners, clubtails, spiketails, and emeralds are among these relatively long-lived species. No California odonate is known with certainty to survive winter as an adult and breed the following spring, although midwinter sightings of the Variegated Meadowhawk suggest that this species might be capable of doing so, at least in some years.

In the final stage—the metamorphosis—the last larval instar leaves the water. In some species it may move some distance from water, even climbing into vegetation. In other species it simply crawls up onto the shore. The exoskeleton along the back splits open, and the adult dragonfly emerges. It is initially somewhat stunted in form and colorless. Blood is pumped into the wings and abdomen to expand them. When the adult form is achieved, usually after a few minutes to over an hour, the young dragonfly, called a teneral, flies away from the emergence site to forage and mature fully. Teneral dragonflies typically have somewhat flimsy, glistening wings, and their bodies are still soft, so they are easily damaged if handled. This is a brief but dangerous time in a dragonfly's life, and many tenerals are preyed upon, for example, by birds and even other dragonflies.

The cast-off larval exoskeleton, called an exuvia, is paper light but often remarkably sturdy. Exuviae of species that emerge in large numbers over a short time may be found littering the shore of lakes or ponds, perched on rocks in streambeds, and hanging from emergent vegetation. The characteristic features of the larvae are preserved in fine detail on exuviae, which often can be identified to genus if not species. They also may indicate some measure of breeding distribution and habitat use by the various species in a region.

The following key will help you identify late-instar larvae (those with definite wing pads) and exuviae to family and, in some cases, to subfamily. Some larvae are especially distinctive, and a few of these are illustrated in fig. 8. However, in some

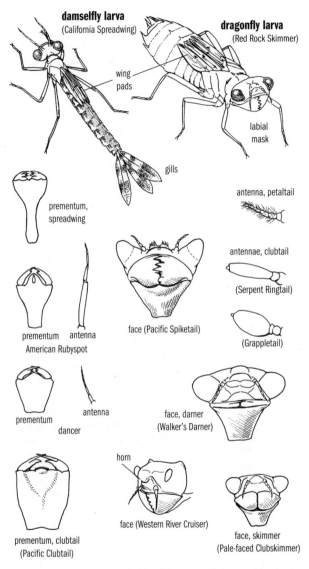

damselfly larva
(California Spreadwing)

dragonfly larva
(Red Rock Skimmer)

wing pads

labial mask

gills

prementum, spreadwing

antenna, petaltail

antennae, clubtail

(Serpent Ringtail)

(Grappletail)

prementum   antenna
American Rubyspot

face (Pacific Spiketail)

prementum   antenna
dancer

face, darner
(Walker's Darner)

horn

prementum, clubtail
(Pacific Clubtail)

face (Western River Cruiser)

face, skimmer
(Pale-faced Clubskimmer)

Figure 8. Key characteristics for identifying dragonfly larvae with regard to family.

groups, most notably the pond damsel family (Coenagrionidae), the differences among species and even genera are difficult to detect, requiring high-powered magnification and often reference to a large series of specimens or some of the sources cited in the "References" section.

Family and Subfamily Key to Dragonfly Larvae

8a. Horn on top of head between antennae; legs long for body size ("spidery") ....
.... cruisers (Libellulidae: Macromiinae)
8b. No central horn atop head ....
.... emeralds (Libellulidae: Corduliinae)
and skimmers (Libellulidae: Libellulinae)

Distribution

Any discussion of the distribution of California's dragonflies must be prefaced with the acknowledgment that a number of species are represented by just a few, scattered records and that large areas of the state have been little explored for odonates. That being said, the broad outlines of statewide distribution can be described for many species, and regional faunas can be defined by examining the patterns of these distributions (figs. 9, 10).

Because of its size and varied topography, California has a diverse dragonfly fauna of 108 species, ranging from subtropical to boreal forms. The group of species around a boggy Sierran lake is quite different from that of a Central Valley marsh, and both differ greatly from that of a Colorado Desert hot spring. The diverse regional landscape compensates somewhat for the fact that the arid climate of much of the state limits the extent and variety of aquatic environments that provide for rich dragonfly faunas in more humid climes. For example, Ohio, a smaller state with less geographic and climatic diversity but wetter summers than California, hosts about 162 species of odonates.

A glance at a map of California (fig. 10) quickly reveals what is well known to most Californians: the state is composed of a series of large and diverse topographic features that run from north to south. From west to east, these include a string of coastal ranges; a large, central valley; high mountains (the Sierra Nevada and Cascade Range); and desert country along the eastern border of the state (the Great Basin in the north and the Mojave and Colorado Deserts to the south).

The north-south orientation of most of the state's mountain ranges aligns them broadside to the prevailing path of winter storms headed east off of the Pacific Ocean. As a result, the mountains have a major impact on the distribution of precipita-

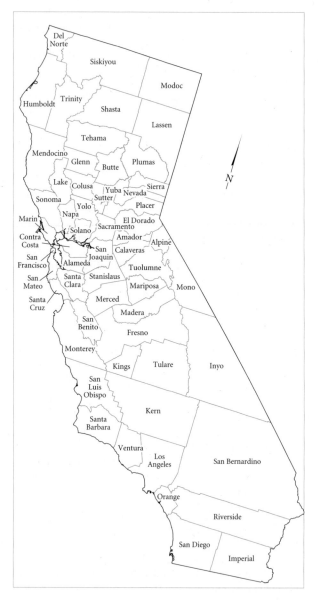

Figure 9. Map of California counties.

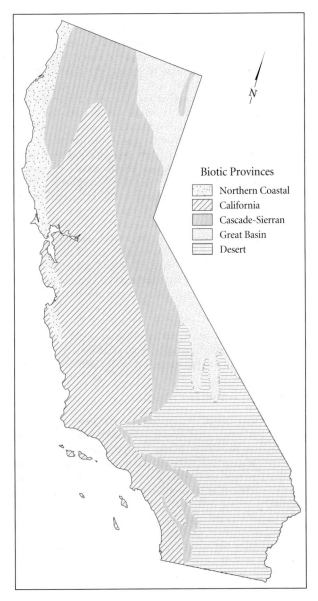

Figure 10. Map of California biotic provinces.

tion in California, most of which is delivered by those winter storms. A good deal of that precipitation is deposited in the form of rain (and some snow, primarily in the north) as moisture-laden air rises up over the western slopes of the Coast Ranges. As the cool air drops down the eastern slopes of the coastal hills, it retains some moisture, limiting precipitation in the Central Valley, especially in the south, to modest proportions. Even more rain and snow are rung out of these Pacific storms as the air rises again over the higher interior ranges; by the time the storms reach the eastern slopes they often have little moisture left to spare, resulting in desert conditions in those areas.

This dramatic disparity in the distribution of water in California has been greatly altered by human activity in the past century. Giant aqueducts carry water from rivers in the north to urban centers in the drier south. Thousands of artificial bodies of water, from immense reservoirs to small stock ponds, have been created in the mountains and foothills to harness runoff that formerly produced extensive seasonal wetlands in the Central Valley. Both the natural distribution of water in the state and its manipulation by humans are significant influences on the distribution of odonates.

Regional dragonfly faunas are perhaps best discussed in the context of biotic provinces, which are large areas primarily defined by topography, climate, and vegetation. The following provinces, and their associated odonate faunas, occur in California.

Northern Coastal Province

The Northern Coastal Province is a region of heavy winter rainfall and cool, foggy summers on the western slope of the Coast Ranges, from the Oregon border to the northern San Francisco Bay Area (where small pockets of this province are found within the next province, south to the Monterey area). This is what Californians think of as "redwood country." Most of this area was originally clothed in humid conifer forest, but much of it has been logged for timber or cleared for agriculture. Aquatic habitats are primarily fast-flowing rivers and creeks bordered by willows and alders but include some marshes bordering coastal lagoons and bays. Some mountain lakes are found, primarily in the north. Only one dragonfly species, the Swift Forktail *(Ischnura erratica),* has its range within the state almost completely restricted to this province. Other species in this region are also found in the Cascade-Sierran Province (primarily at higher elevations and to the north) and the California Province (primarily at lower elevations and to the south).

California Province

The California Province is the heart of California and includes the Central Valley, the inner Coast Ranges (and valleys such as the Napa Valley) from approximately Lake and Mendocino Counties southward to the San Francisco Bay Area, and all the coastal districts from there southward to the Mexican border (except for the higher mountains of southern California). Few odonate species have been recorded from the Channel Islands, and all are species found along the nearby mainland, so those islands are best included in this province.

Although topographically varied, the various parts of this region share a Mediterranean climate (cool, wet winters and hot, dry summers). Except for the major river systems and large flood basins of the Central Valley, which receive runoff from the Sierra Nevada snowpack through spring and into summer, natural wetlands in this province historically tended to be seasonal, containing little standing water by late summer. However, a large percentage of the region's natural wetlands have been drained, developed, or altered to a great extent. The flow of water through many of the major rivers is regulated by dams, resulting in higher, colder water levels later in summer than in the past. The creation of reservoirs has provided large areas of permanent open water in areas that originally had none. Aqueducts and irrigation canals move large quantities of water over long distances from one part of the province to another. All of this activity has no doubt had a profound affect on odonate populations.

This province, which is restricted to California and part of adjacent Baja California, contains the greatest number of species of plants and animals endemic to California (i. e., not found elsewhere) of any province in the state. Not surprisingly, this includes our only endemic odonate species (the San Francisco Forktail *[Ischnura gemina]* and Exclamation Damsel *[Zoniagrion exclamationis]*). A few other species (e.g., the Lavender Dancer *[Argia hinei]* and Serpent Ringtail *[Erpetogomphus lampropeltis]*) are known only from this province within California (although they also occur outside the state), and several others (the California Spreadwing *[Archilestes californica]*, Great Spreadwing *[Archilestes grandis]*, Black Spreadwing *[Lestes stultus]*, Sooty Dancer *[Argia lugens]*, Walker's Darner *[Aeshna walkeri]*, Bison Snaketail *[Ophiogomphus bison]*, Pale-faced Clubskimmer *[Brechmorhoga mendax]*, and Red Rock Skimmer *[Paltothemis lineatipes]*) are characteristic of this province and relatively scarce outside of it within the state. Many other species, however, are shared with the Desert Province.

Cascade-Sierran Province

The Cascade-Sierran Province includes all the high-mountain areas, dominated by mixed-conifer forests, throughout the state—the Klamath Mountains, North Coast Ranges, Cascade Range, Warner Mountains, Sierra Nevada, and higher peaks of some of the small mountain ranges in southern California. These areas are characterized by cold winters, in which some of the precipitation falls as snow, and mild summers, with some precipitation in the form of thundershowers. Still waters at the higher elevations may be frozen during at least part of winter. The dragonfly season begins later at higher elevations, and many species in this province do not emerge in numbers until July through September, well after the peaks of emergence for most lowland populations.

This province contains a great variety of aquatic habitats, both still and flowing waters, including rivers, creeks, springs, seeps, bogs, marshes, wet meadows, and lakes of many sizes and kinds. Such habitat diversity accommodates a rich dragonfly fauna, which is dominated by spreadwings, bluets, darners, emeralds, whitefaces, and meadowhawks. Many of these species are found exclusively, or almost so, within this province in California but are distributed much more widely to the north in the mountain forests of the Pacific Northwest and the boreal forests of Canada. These include the Emerald Spreadwing *(Lestes dryas)*, Taiga Bluet *(Coenagrion resolutum)*, Sedge Sprite *(Nehalennia irene)*, Black Petaltail *(Tanypteryx hageni)*, Canada Darner *(Aeshna canadensis)*, American Emerald *(Cordulia shurtleffii)*, Mountain Emerald *(Somatochlora semicircularis)*, Ringed Emerald *(Somatochlora albicincta)*, Spiny Baskettail *(Tetragoneuria spinigera)*, Four-spotted Skimmer *(Libellula quadrimaculata)*,

Chalk-fronted Corporal *(Ladona julia)*, Hudsonian Whiteface *(Leucorrhinia hudsonica)*, Red-waisted Whiteface *(Leucorrhinia proxima)*, Crimson-ringed Whiteface *(Leucorrhinia glacialis)*, and Black Meadowhawk *(Sympetrum danae)*.

Great Basin Province

The Great Basin Province is "sagebrush country," the area of high desert with cold winters, dominated by sagebrush-covered flats and juniper-clad hills east of the Cascade Range and Sierra Nevada. Dragonfly habitats in this region include small rivers and creeks; isolated springs, both hot and cold; freshwater marshes; and ponds and lakes, many of which are intermittent and alkaline. Species primarily restricted to this province within California include the River Bluet *(Enallagma anna)*, Alkali Bluet *(Enallagma clausum)*, and Pale Snaketail *(Ophiogomphus severus)*. Some other species characteristic of the region, but also found in the southern Desert Province, are the Paiute Dancer *(Argia alberta)*, Brimstone Clubtail *(Stylurus intricatus)*, Bleached Skimmer *(Libellula composita)*, and Desert Whitetail *(Plathemis subornata)*. In addition, a number of species of the Great Basin—the River Jewelwing *(Calopteryx aequabilis)*, Common Spreadwing *(Lestes disjunctus)*, Lyre-tipped Spreadwing *(Lestes unguiculatus)*, Western Red Damsel *(Amphiagrion abbreviatum)*, Great Basin Snaketail *(Ophiogomphus morrisoni)*, Dot-tailed Whiteface *(Leucorrhinia intacta)*, and a variety of meadowhawks—are shared with the Cascade-Sierran province.

Desert Province

The Desert Province includes both the Mojave and Colorado Deserts, which have similar dragonfly faunas and also share many species with the Great Basin and California Provinces. Natural dragonfly habitats of the region are primarily spring runs and alkaline basins, but also desert rivers such as the Mojave and the Colorado. Irrigation projects, especially those involving the Colorado River, have significantly altered the distribution of water in this region and created a number of nonnative aquatic habitats, such as irrigation canals and artificial lakes and rivers. The most dramatic of these is the Salton Sea and environs, inadvertently flooded in the early part of the twentieth century.

Species characteristic of this province in California are the Powdered Dancer *(Argia moesta)*, Blue-ringed Dancer *(Argia sedula)*, Double-striped Bluet *(Enallagma basidens)*, Desert Forktail *(Ischnura barberi)*, Citrine Forktail *(Ischnura hastata)*, Rambur's Forktail *(Ischnura ramburii)*, Russet-tipped Clubtail *(Stylurus plagiatus)*, Red-tailed Pennant *(Brachymesia furcata)*, Marl Pennant *(Macrodiplax balteata)*, Roseate Skimmer *(Orthemis ferruginea)*, and Mexican Amberwing *(Perithemis intensa)*. A number of these species were detected in the state only within the past few decades and appear to have spread into this region as a result of extensive human alterations of aquatic systems.

Watching Dragonflies

You can easily watch and enjoy many species of dragonfly with the naked eye. However, you will need a few aids to help you accurately identify most species.

A pair of binoculars is very useful. The closer they can focus, the better. Many dragonflies, if approached cautiously, will allow you to get within a few feet. The real trick is to spot them before they see you—and they can see you coming from a good ways off! By carefully approaching a perched dragonfly from behind, you increase your chances of getting close to it.

A spotting scope can also be useful for watching dragonflies. Many species that forage while constantly in flight, such as darners and gliders, are apparently resting between foraging bouts when they go to perch and will often stay perched in the same spot for some time. You should watch for the distinctive behavior of these types of foragers when seeking a roost. The dragonfly will fly in close to vegetation as if looking for a perch, and then back off a few feet. It may repeat this behavior a number of times before landing, so stand still and be patient. Once the animal lands, give it a minute or two to settle down. It may become quite unwary and allow close approach. If you have a spotting scope, get as close to the dragonfly as the focusing on your scope will allow, then study it at leisure. Even very tiny features (such as the tubercle on the undersurface of the first abdominal segment of some species of *Aeshna*) occasionally can be observed in this way.

However, in many situations, particularly those involving

small species (e.g., most damselflies) or very similar looking species (such as the baskettails *[Tetragoneuria]*), you will need a net with which to capture the animal, and some sort of hand lens with which to examine it; 10 to 12× (or higher) magnification is useful. In a pinch you can use your binoculars, turning them around and looking through them backward. A portable source of light may come in handy.

A good dragonfly net has a large mouth (at least 30 cm [1 ft] in diameter, but preferably larger) and a long handle (at least 1 m [3 ft] in length) and is constructed of a light material (e.g., wood, aluminum, bamboo). The mesh should be fine but sturdy. You may build your own net or purchase one (see the reference list at the end of the book).

When netting perched dragonflies, approach from behind. With a full, swift downward or sideways stroke, swing right through the animal, centering it in the net mouth so that it is funneled to the bottom of the net bag, then quickly flick your wrist upward to loop the bag across the rim of the net. Attempting to net dragonflies in flight is much trickier but often the only way possible to catch species that seldom perch within net range or seem to be constantly foraging on the wing. The approach, swing, and follow-through are similar and best taken from behind and below. If possible, watch a dragonfly for a while before trying to net it; if it is following a regular foraging beat, you may be able to strategically position yourself for a successful swing. Or it may perch and allow you to observe, photograph, or sketch it before you net it to make sure of your identification.

A smaller net, like those used for taking fish out of aquaria but with a long handle (about a half meter [1.5 ft]), can be dropped quickly over dragonflies on the ground or low in vegetation.

Removing a dragonfly from a net is usually not too difficult. Try keeping the animal in a small pocket of the net bag to prevent it from flying about and damaging itself or escaping. Carefully insert your hand into the net, and gently grasp the wings. Unlike butterfly wings, dragonfly wings are sturdy and can withstand gentle handling, with no harm to the animal. Hold the animal by all four wings folded together over the back. Don't worry about getting bitten. A large dragonfly may occasionally nip at you, but such a nip is usually more startling than anything else and is unlikely to draw blood or cause much pain.

Although a net will be essential for catching and identifying all the species you encounter, it is not the only tool you can use to study dragonflies. You may prefer to use a camera—for film, digital imagery, or videotape. It is also useful to write down your observations in a notebook or make sketches of what you see.

The purpose of this guide is to help you identify dragonflies in the field without the need to collect them, although it will be necessary to examine certain species in the hand, especially when first learning to recognize them, to be sure of your identifications. For information on collecting and preserving specimens, consult the references at the back of the book.

Identifying Dragonflies

The first thing most people notice about dragonflies is their color. However, coloration by itself is usually insufficient to allow identification of most dragonflies as to species; for example, fully adult males of most of our bluets and dancers are blue and black. In addition, color can vary quite dramatically within species as a result of age, sex, geographic variation, and even temperature. Teneral dragonflies, within a few hours after emerging, often lack the adult colors and are essentially a dull tan or pale olive color. Males of many species often do not attain full adult coloration until they become reproductively active, which may take several days, and during the interim they may be colored much like females. Females may also change color with age, or they may occur in two different color phases, one different—and typically less colorful—than the male pattern (gynomorphic), and one similar to the color of males (andromorphic). In particular, females of some species (e.g., some damselflies and darners) can show a particularly complicated pattern of variation in color, occurring in two or more color phases and also changing color with age or temperature.

The wings of young adults appear bright and shiny, whereas the wings of old dragonflies typically are dull, somewhat opaque, and tattered. Breeding activity may also affect coloration: females that have oviposited in soil may have the tips of their abdomens coated with dried mud. In species in which the males become pruinose on the abdomen, the pruinescense on segments 5 through 7 is often rubbed off by the female's feet where she grasps the abdomen while in the wheel position.

Ambient temperature can affect color in species such as bluets, dancers, and darners that have blue markings (excluding blue pruinescense). The blue areas on these animals may fade when the temperature drops, becoming dark slate or dull purple, and then revert to bright blue when the temperature rises. This reversible color change can occur within minutes if the animal is rapidly chilled or warmed.

For these reasons, you can't rely on color alone to identify species of dragonflies. More important aspects to note are pattern and structure. Many species have distinctive patterns of dark (typically black or brown, perhaps with some iridescence) and light (white, yellow, blue, green, red) on the thorax and abdomen. Some species have distinctive patches of color on the wings. Differences in these patterns among species may be dramatic or subtle, but they tend to be fairly constant and usually can be viewed through binoculars at close range (if the dragonfly sits still!). Even in species that become pruinose or exhibit temperature-dependent color changes that can obscure these markings, the "ghost" of a pattern can often be seen upon close examination.

Structural features are often critical for identification, especially in large groups of similar species, such as the bluets. Often the most species-specific structures on dragonflies are those that establish the very precise linkage between the sexes during mating. In males, these include the abdominal appendages, used to grasp the female, and the secondary genitalia, under the second abdominal segment. In females, these may include the abdominal appendages and genitalia, the prothorax and plates on the front of the pterothorax of damselflies, or the back of the head in some typical dragonflies. Some of these features can be observed in the field at close range with binoculars, but more often than not individuals need to be captured and examined in the hand.

Overall appearance and behavior are often very useful indicators of the general group to which a particular dragonfly belongs. For example, a medium-sized odonate with a damselfly body type (long, thin abdomen and widely spaced eyes) that holds its wings spread to the side when perched is in all likelihood a species of spreadwing, in either the genus *Lestes* or the genus *Archilestes*. Likewise, a small, mostly black dragonfly with a bright white face is recognizable as a species of whiteface, in the genus *Leucorrhinia*.

When using this book to identify an unfamiliar dragonfly by sight, first, use the family key and fig. 11 to narrow your choices.

Once you become familiar with odonate families, you will be able to skip the key in most cases. Second, skim through the illustrations of species within the family you have selected and look for species similar to your unknown animal. Third, read the descriptions of possible candidates, as well as the section on similar species. Fourth, check the distribution of your likely candidates. At this point you should have narrowed down the possibilities sufficiently to make a confident identification in many cases, but sometimes (especially with damselflies) it will be necessary to capture the animal and examine small, structural features in order to be positive.

Species are described and illustrated as they appear in California. Some species vary in appearance in different parts of their range, so caution is warranted in using this guide well outside the state, especially in the eastern half of North America. The descriptions are useful throughout much of the northwestern part of the continent, however.

Taxonomy and Nomenclature

Both scientific and English names are provided for all species discussed in this guide. The binomial (two-name) nomenclature used by scientists to describe species includes the generic name, which is capitalized (a genus typically is a set of related species, but may be monotypic, i.e., contain only one species), and the specific epithet, which is an adjective or noun that is not capitalized (even if based on a proper name).

The scientific names are Greek, Latin, or latinized English. For example, *Libellula quadrimaculata* is a species in the genus *Libellula*. The specific epithet means *four-spotted*, in reference to the small spots at the nodus of each wing in this species. In a few instances, a third name is used to describe a recognizable subspecies (e.g., *Ophiogomphus morrisoni nevadensis*, the paler form of *Ophiogomphus morrisoni* found in the desert country of the Great Basin). The scientific nomenclature used in this guide follows that of the most recent handbooks of North American damselflies (Westfall and May 1996) and dragonflies (Needham, Westfall, and May 2000).

In this guide, I have used the English names for North American odonates that have recently been standardized by the Dragonfly Society of the Americas (Paulson and Dunkle 1999). To a great extent, these names have also been constructed to have both a group (i.e., generic) component and a specific component. Thus, all the species in the genus *Libellula* (often referred to as the king skimmers) are called skimmers; the English name given to *Libellula quadrimaculata* is Four-spotted Skimmer. Not all English names are this closely linked to their corresponding scientific names, and in some cases the English group name is not tied exclusively to a particular genus (e.g., the Roseate Skimmer is in the genus *Orthemis*, not *Libellula*), but they still remain useful for categorizing species into readily identifiable groups such as skimmers, darners, bluets, meadowhawks and so forth. The English names are not common names, that is, most do not have a history of widespread usage. Indeed, they are of such recent vintage that they are not that familiar even to all professional odonatologists.

Both English and scientific names are useful in certain circumstances. English names are more readily assimilated by English speakers and do not require translation, whereas scientific names are used universally by biologists. It is perhaps most useful to learn both names and use whichever seems appropriate.

The species composition and order of the various families within each suborder are also based on the handbooks referred to above and agree with those most widely accepted by North American odonatologists. However, the relationships of genera and species within families are more controversial. Indeed, a common practice of odonatologists has been to avoid adoption of any particular taxonomic arrangement and simply list genera, and species within genera, in alphabetical order. In this guide, genera and species have been ordered to facilitate field identification as much as possible. Often this means that closely related species are grouped together, but not always. As a result, you should not attempt to interpret the order of the species listed as a "family tree."

About the Maps

The best range maps are dot maps, which plot known locations as individual points. Dot maps, however, are not suited for publication at the small size required for the range maps in this guide.

The range maps provided here were created by first plotting known locations on a map of the state, encircling the appropriate cluster(s) of plotted points, and finally, shading in the enclosed area(s). In some cases, isolated locations were plotted as single points.

Determination of boundaries of each distribution was based on knowledge of topography and presence or absence of suitable habitat. It should not be assumed, however, that a species is necessarily evenly dispersed throughout the range shown. For example, the distribution of most odonates in the desert regions of the state is extremely sparse because of the limited availability of water. The maps, then, should be considered rough guides to general range and not detailed depictions of known distribution.

Family and Subfamily Key to Adult Dragonflies

1a. Delicate, slender bodied; compound eyes somewhat stalked and widely separated; fore and hind wings of similar shape and size, typically held together over the abdomen (partly spread in one family). 2

1b. Heavy bodied; large compound eyes touch each other or are relatively close together; hind wings broader based than fore wings, wings held spread out to sides 4

    2a. Wings with patches of black or brown at tips or red or brown at base, not constricted near base . . . . . .
    . . . . broad-winged damsels (Calopterygidae)

    2b. Wings constricted near base (petiolate), without colored patches. 3

3a. Wings typically held partly spread out to sides; pterostigma relatively long, equal in length to width of eye
. . . . spreadwings (Lestidae)

3b. Wings typically held together over abdomen; pterostigma about as long as wide, smaller than width of eye . . . .
. . . . pond damsels (Coenagrionidae)

    4a. Compound eyes well separated. 5

    4b. Compound eyes touching, if only slightly 6

5a. Pterostigma long and narrow; body usually black with

spreadwing

broad-winged damsel

pond damsel

spreadwing pterostigma

pond damsel pterostigma

petaltail head (top view)

petaltail pterostigma

spiketail head (top view)

clubtail pterostigma

darner head (top view)

club-shaped anal loop
in hind wing of an emerald

skimmer head (top view)

foot-shaped anal loop
in hind wing of a skimmer

Figure 11. Key characteristics for identifying adult dragonflies with regard to family.

yellow markings; found at hillside seeps and mountain bogs . . . . Black Petaltail (Petaluridae)

5b. Pterostigma relatively thick and oval shaped; variably patterned; usually found along rivers and streams (some at lakes) . . . . clubtails (Gomphidae)

    6a. Eyes just touching; large, black body marked with yellow . . . . Pacific Spiketail (Cordulegastridae)

    6b. Eyes broadly contiguous; variously colored 7

7a. Large (>55 mm in length); robust thorax either entirely green or brown striped with blue, green, or yellow; long, cylindrical abdomen either extensively blue or purple, or with patchwork of blue, brown, or yellow spots; females use ovipositor to insert eggs in vegetation, logs, mud, etc . . . . darners (Aeshnidae)

7b. Variable in size, shape, and color, but not as described above; females deposit eggs in water (some on ground or vegetation) 8

    8a. Large; black or dark brown with yellow saddles on top of abdominal segments 3 through 8, abdomen of male club shaped at tip; single pale yellow stripe on side of pterothorax; fast fliers patrolling streams and rivers or feeding over roads and clearings . . . . Western River Cruiser (Libellulidae: Macromiinae)

    8b. Not as above 9

9a. Anal loop club shaped; abdomen black or black with row of orange yellow patches along side of spindle-shaped abdomen; eyes glowing emerald green or turquoise; hang vertically when perched; found at wooded ponds, lakes, or boggy meadows . . . . emeralds (Libellulidae: Corduliinae)

9b. Not as above; anal loop foot shaped; color, size, and pattern variable, many with patches of color on wings; occupy a wide range of habitats and include some of the most familiar species . . . . skimmers (Libellulidae: Libellulinae)

SPECIES ACCOUNTS

Each species account begins with the total length and wing span (wings spread, as in flight) provided to the nearest quarter centimeter and to the nearest half inch. Usually an average or range is given, not as a precise measurement but as an indication of relative size. You may find it easier to make a rough estimate of the size of free-flying individuals in terms of inches, but metric units may prove more useful for in-hand measurements especially of small structural features. Remember that for many species, smaller individuals are generally encountered later in the flight season.

After these dimensions, each account provides a description of the species, with tips for distinguishing it from similar species. Behaviors that may aid in finding or identifying the species follow. Each account ends with information on geographic distribution, habitat, and the extent of the flight season.

BROAD-WINGED DAMSELS
(Calopterygidae)

These large, colorful stream dwellers are among our most attractive and easily identified damselflies. Two genera occur in North America, one species of each in California. The wings are fairly broad—not constricted basally as on our other damselflies—and have a dense network of veins. Colored patches in the wings, which are especially bright in males, are exposed in territorial defense and courtship behaviors (wing flicking and display flights) and give rise to the English names *jewelwing* and *rubyspot*. Their flight is somewhat slow and butterflylike.

Jewelwings *(Calopteryx)*

A large, cosmopolitan genus, jewelwings are represented in California by one species. Four other species occur in North America, all east of the Rocky Mountains.

RIVER JEWELWING *Calopteryx aequabilis*
Pl. 1

LENGTH: 4.5 to 5 cm (2 in.); **WING SPAN:** 6.5 to 7 cm (2.5 to 3 in.)

DESCRIPTION: This is an unmistakable creature—a large damselfly with long legs and a brilliant metallic body. The male is emerald green, the female bronze green. Dark wing tips, glossy black on the male and dark brown on the female, occupy approximately the outer fourth of the fore wings and nearly the outer half of the hind wings. The pterostigma on the female's wing is a striking white, and the inner wing is smoky.

SIMILAR SPECIES: We have no other damselfly like it. The closest is the American Rubyspot *(Hetaerina americana),* which has red or brown wing bases and clear wing tips.

BEHAVIOR: Breeding males on territories occupy exposed perches on low vegetation over water. They actively engage in skirmishes over preferred perches and potential mates. Males guard their ovipositing mates, but not in tandem. Nonbreeding individuals do not wander far, but may be found foraging in openings in riparian thickets near breeding sites.

DISTRIBUTION: This beautiful species is known from a few rivers— the Eel, Mad, Van Duzen, Klamath, Shasta, Pit, and Susan—in northern California from Mendocino and Lassen Counties northward. Recorded elevations range from near sea level in Humboldt County to over 2,100 m (7,000 ft) in the North Coast Ranges.

HABITAT: The River Jewelwing frequents the heavily vegetated edges of large streams and rivers. It is often found where dense, herbaceous growth such as reeds, grasses, sedges, and loosestrife extends up to and beyond the water's edge.

FLIGHT SEASON: It has been found on the wing in the state from May through July, but most frequently after mid-June.

Rubyspots *(Hetaerina)*

This is a large genus of primarily Neotropical damsels. A few species are found in North America, the most common and widespread of which reaches California.

AMERICAN RUBYSPOT   *Hetaerina americana*
Pl. 1, Fig. 8

LENGTH: 4 to 4.5 cm (1.5 in.); **WING SPAN:** 6.5 cm (2.5 in.)

DESCRIPTION: The male is a large and showy damselfly. It has dark upper parts with metallic reddish purple iridescence and dark wing bases that are bright red on the upper surfaces. In California, the male usually lacks pterostigmata in the wings, whereas the female has tiny whitish ones. The female is dark purplish brown above with pale stripes on the sides of the thorax and has dull brown or amber wing bases.

SIMILAR SPECIES: None of our other damselflies are like it. The most similar is the River Jewelwing *(Calopteryx aequabilis)*.

BEHAVIOR: Males congregate at suitable oviposition sites along streams and flick their wings to attract females. Foraging individuals do not wander far from riparian corridors. Females completely submerge to oviposit in aquatic vegetation.

DISTRIBUTION: This damselfly is found statewide from sea level to approximately 1,500 m (5,000 ft).

HABITAT: The American Rubyspot occupies a wide range of flowing waters, from small spring runs and irrigation ditches to larger streams and rivers.

FLIGHT SEASON: The season is fairly long, especially at lower elevations, lasting from February through November.

SPREADWINGS (Lestidae)

Spreadwings are distinctive, medium-sized to large damselflies that hold their wings spread out to the sides, somewhat like dragonflies, when perched. Other damselflies may spread their wings, and spreadwings may bring their wings together on occasion, but only briefly and not habitually. Oviposition is accomplished in tandem. Females oviposit in vegetation, often some distance above water level or even over dry pond beds. Eggs hatch in spring when water levels rise. The seven species of spreadwings found in California include two large, stream-dwelling species and five smaller, pond-dwelling species.

Stream Spreadwings *(Archilestes)*

Two of the six species in the primarily Neotropical genus *Archilestes* reach the United States; both occur in California. These are large and robust damselflies. Although not brightly marked, both our species have a pale stripe on the side of the thorax, and this feature plus large size and spread wings makes them relatively easy to identify. In September and October, the usual peak of breeding activity in much of the state, they are characteristic denizens of willow-bordered pools of rocky streams in foothill and lower-montane canyons.

CALIFORNIA SPREADWING　　　*Archilestes californica*
Pl. 2, Fig. 8
LENGTH: 4.5 to 6 cm (2 to 2.5 in.); **WING SPAN:** 5.5 to 7 cm (2 to 3 in.)
DESCRIPTION: This is a large, stocky spreadwing that in the field looks predominantly brown. Most of the dorsal and lateral surfaces are brown, with some patches of dull black on the thorax. The middle abdominal segments may exhibit some green iridescence. In life, the pterostigma is light brown. The mature male and the female look alike, except that the male has bright blue eyes above and develops gray pruinescence on abdominal segments 9 and 10. The most distinctive field mark is an incomplete whitish stripe along the side of the thorax that fades out as it approaches the base of the hind wing.
SIMILAR SPECIES: Pond spreadwings (*Lestes*) are smaller, differ-

ently colored, and usually found at ponds. The only species easily confused with the California Spreadwing is the Great Spreadwing *(A. grandis)*, which is a bit larger on average, has a darker thorax above (predominantly blackish with green iridescence), and has a complete, yellowish side stripe. The pterostigma of the mature male Great Spreadwing is blackish versus the tan of a California Spreadwing. In hand, the paraprocts of a male California Spreadwing are aligned in parallel (they may temporarily be spread on occasion) versus the diverging paraprocts of a male Great Spreadwing.

BEHAVIOR: These are quick and agile fly catchers, darting out to catch prey and then returning to perch. They are prone to fly up and into vegetation when alarmed. Prior to breeding, they may wander some distance from water, appearing in foothill brushland and suburban yards. When you first encounter them, their ovipositing behavior is rather startling. Tandem pairs crawl slowly over the surfaces of willow or alder stems about 0.5 to 1 cm (0.25 in.) in diameter. The female, drawing her abdomen up between her legs, thrusts her ovipositor into the branch and inserts eggs, one by one, in little clusters, into the cambium layer. The pair then backs down along the stem a few steps and repeats the process. The result is a series of small scars along the length of the stem. Oviposition occurs in fall, when water levels are typically at their lowest, and oviposition sites may be 1 to 2 m (3 to 6 ft) above what little water, if any, remains in the stagnant pools of intermittent streams. Ovipositing pairs are thus often at or near eye level and easily observed at close range. They are sometimes unwary enough to be picked up by hand. Because oviposition sites do not appear to be randomly distributed along a stream course, large numbers of tandem pairs can be found ovipositing concurrently at select sites.

DISTRIBUTION: Found in scattered locations from Washington and Idaho southward to Baja California, this spreadwing occurs primarily within the California Province in California, west of the Pacific Crest and the southern deserts. It ranges from sea level to about 1,200 m (4,000 ft) but is most typically found in foothill canyons of the Coast Ranges from Siskiyou County southward and along the western slope of the Sierra Nevada. It has recently been found east of the Sierra Nevada along the Susan River in Susanville.

HABITAT: The California Spreadwing is found in the vicinity of small, often intermittent, streams and the backwaters of larger streams and rivers. Breeding habitat consists of pools, bordered with willows and alders, within the streambed, or occasionally, ponds. Foraging individuals not engaged in breeding activity may be found some distance from water in open woodland, chaparral, and weedy areas.

FLIGHT SEASON: Emergence may begin as early as the third week of June, but the mature adult form of the California Spreadwing is most often seen at breeding sites in September and October. It often flies into the first half of November.

GREAT SPREADWING   *Archilestes grandis*
Pl. 2

LENGTH: 5 to 6.25 cm (2 to 2.5 in.); **WING SPAN:** 6.5 to 8 cm (2.5 to 3 in.)

DESCRIPTION: California's largest damselfly is an impressive species that resembles the California Spreadwing *(A. californica)* in that the male has bright blue eyes above and gray pruinescence on the terminal abdominal segments, but it has a dark, mostly black thorax with metallic green highlights above and a complete yellowish side stripe. The middle abdominal segments are iridescent green above. The pterostigma is nearly black on the mature male, paler on the female and the teneral individual. The paraprocts of the male are divergent rather than parallel as in the California Spreadwing.

SIMILAR SPECIES: See the California Spreadwing.

BEHAVIOR: Great Spreadwings behave much like California Spreadwings. Pairs oviposit in tandem, the female inserting eggs in the leaf petioles of trees and stems of shrubs from one to several meters (3 ft or more) above the surface of pools in the stream. After hatching, the larvae "jump" to the water below.

DISTRIBUTION: The Great Spreadwing is known from scattered locations in southern California west of the deserts, as well as the Coast Ranges from Trinity County southward to the San Francisco Bay Area and along the western slope of the Sierra Nevada. The elevation range is similar to that of the California Spreadwing.

HABITAT: The Great Spreadwing occurs in much the same habitats as does the California Spreadwing but seems more restricted to flowing water. The distributions of the two species in the state broadly overlap, and they coexist at a number of sites.

FLIGHT SEASON: Known flight dates in the state range from late July to mid-January, with the majority of records in September and October.

Pond Spreadwings *(Lestes)*

This is a large genus with a worldwide distribution. Four of our five species are relatively common and widely distributed in North America; the fifth is nearly endemic to California. The greatest diversity of species in California is found in the northeastern part of the state. They breed in boggy meadows, ponds, and small lakes, often temporary ones, and occasionally along slow streams. All are rather similar in appearance and fairly easily distinguished from other damselflies (except the even larger stream spreadwings *[Archilestes]*) by their relatively large size and characteristic stance when perched: abdomen hanging down and wings spread apart. The eyes of mature males are blue; females and immature males often have paler blue, gray, or brown eyes. Abdominal segments 9 and 10 develop gray pruinescence, producing a pale-tipped abdomen.

SPOTTED SPREADWING   *Lestes congener*
Pls. 3, 4

LENGTH: 3.5 to 4 cm (1.5 in.); **WING SPAN:** 4.5 cm (2 in.)

DESCRIPTION: This species has the typical *Lestes* pattern. The adult is blackish above and whitish below. The top of the thorax is black and striped with a very thin central and two thin lateral pale green or tan lines, and the thorax may become pruinose with age. The English name derives from the presence of four black spots—two on either side—of the pale undersurface of the thorax (usually visible only in the hand). The short and stubby paraprocts of the male are distinctive. The pterostigma is dark brown on an adult, pale on a teneral individual. A fresh teneral is bronzy, washed with salmon pink.

SIMILAR SPECIES: The most similar species is the Black Spreadwing *(L. stultus)*, which is also blackish above and has similar black spots in the pale lower areas of the thorax. The two species are often found together in the Central Valley and surrounding foothills. The Black Spreadwing is larger and stockier, and its abdomen is more noticeably metallic green above. However, individuals should be netted and examined in hand for certain identification. The male's paraprocts are shorter and stubbier than those of other *Lestes* species. The ovipositor of the female Spotted Spreadwing is shorter than abdominal segment 7 (longer than the same segment on the Black Spreadwing and the Emerald Spreadwing *[L. dryas]*).

BEHAVIOR: At breeding sites, these spreadwings are rather unwary, flying slowly and low over and through patches of emergent vegetation, frequently stopping to perch. Pairs oviposit in tandem in emergent vegetation, apparently preferring dry stems. The overwintering eggs are highly resistant to desiccation. Nonbreeding individuals may wander far from water, even into suburban yards.

DISTRIBUTION: The Spotted Spreadwing is more widely distributed in the state than any other spreadwing, being found in nearly every biotic province except the southern deserts. It might be found around any temporary pond, roadside ditch, or freshwater marsh from the Oregon border to San Diego County and from sea level to over 2,100 m (7,000 ft) in the Sierra Nevada.

HABITAT: Breeding habitat is low, emergent vegetation (sedges, rushes) bordering ponds, bogs, lakes, and occasionally streams.

FLIGHT SEASON: This spreadwing flies from May through early November, beginning earlier at lower elevations.

BLACK SPREADWING   *Lestes stultus*

Pls. 3, 4

LENGTH: 3.5 to 4.5 cm (1.5 to 2 in.); **WING SPAN:** 5 cm (2 in.)

DESCRIPTION: This is our largest and stockiest pond spreadwing. Its thorax pattern is similar to that of the Spotted Spreadwing (*L. congener*), but the dorsal black areas are iridescent with purple and bronze. The dark dorsal areas of the abdomen are usually iridescent emerald green. An immature individual, particularly the female, may have dull green iridescence on the thorax. The pterostigma is dark, nearly black, on the adult and pale on the teneral. The inferior appendages (paraprocts) of the male are foot shaped, the "toes" pointing in. The female's ovipositor is longer than abdominal segment 7.

SIMILAR SPECIES: Recent genetic studies suggest that the Black Spreadwing is merely a well-marked subspecies of the Emerald Spreadwing (*L. dryas*). It is typically larger than the Emerald Spreadwing, with which it has little if any range overlap, and lacks metallic emerald or blue green iridescence on the thorax; but the two forms are otherwise virtually identical. Some immature individuals and females may be impossible to distinguish. The adult is most similar to the smaller and more delicate Spotted Spreadwing, the male of which has small, stubby paraprocts and the female of which has a shorter ovipositor.

BEHAVIOR: Black Spreadwings behave much like Spotted Spreadwings but seem a bit bolder and quicker than Spotteds, an impression perhaps based on their greater size. Those not breeding, especially females, may be found some distance from water in yards and along roadsides.

DISTRIBUTION: This species is nearly confined to California (there are recent reports from southwestern Oregon). Its primary range is in the valleys and foothills from Del Norte County southward to Santa Clara and Stanislaus Counties, but the recent discovery of populations in interior San Diego and San Benito Counties suggests it should be looked for elsewhere in southern California west of the deserts. It is typically found at elevations below 600 m (2,000 ft) in the California Province but has been taken near 1,300 m (4,300 ft) in San Diego County.

HABITAT: Breeding habitat is similar to that of the Spotted and other spreadwings, usually temporary ponds and marshes.

FLIGHT SEASON: An early flyer, it first appears in April, becoming rare by the end of July, but it is occasionally found as late as September. Activity peaks in May and June.

EMERALD SPREADWING   *Lestes dryas*

Pls. 3, 4

LENGTH: 3 to 4 cm (1 to 1.5 in.); **WING SPAN:** 3.5 to 4.5 cm (1.5 to 2 in.)

DESCRIPTION: Stocky and robust for a spreadwing, this species is usually easily identified by the striking metallic emerald or blue green color on top of its thorax and abdomen. The female, particularly if immature, may be duller bronzy green above, and the teneral is bronze and tan. The abdominal appendages are like those of the Black Spreadwing *(L. stultus).*

SIMILAR SPECIES: The larger Black Spreadwing is similar in proportions and morphology, but it is not metallic emerald on the thorax and is typically found at lower elevations. Some females of these two species may be difficult to distinguish. Otherwise, the only other damselfly in our area with similar color on the thorax is the tiny, slender, and very local Sedge Sprite *(Nehalennia irene),* which is not likely to be confused with a stocky spreadwing.

BEHAVIOR: Emerald Spreadwings behave like other spreadwings. Small, high-elevation sites may experience a brief flurry of emergence and rapid breeding because of the short summer season and this species' preference for temporary waters. Away from breeding sites, they may be found foraging low in brushy clearings in open conifer forests.

DISTRIBUTION: This common species is widely distributed in North America and Eurasia. In California it occurs generally at higher elevations, from 400 to 2,400 m (1,300 to 8,000 ft) in the Cascade-Sierran and Great Basin Provinces. There are specimen records from the North Coast Ranges southward to Mendocino and Glenn Counties and from the San Bernardino Mountains.

HABITAT: Breeding habitats include snow ponds, freshwater marshes, boggy meadows, springs, and other (typically small and ephemeral) bodies of water with fairly dense, low emergent vegetation.

FLIGHT SEASON: Most typically found flying June through August and occasionally into September, it may be found as early as late April at lower elevations.

COMMON SPREADWING   *Lestes disjunctus*

Pls. 3, 4

LENGTH: 3.5 to 4 cm (1.5 in.); **WING SPAN:** 4.5 cm (2 in.)

DESCRIPTION: The mature male is a distinctive, frosty blue gray above as a result of heavy pruinescence, especially on the thorax. The paraprocts are long and straight (although sometimes they touch at the tips to form a V or are crossed to form an X). The female and the immature male resemble other spreadwings. The pale bluish lateral stripes on the dark thorax are fairly prominent, and the pale underside of the thorax has at most one small, dark spot on the side near the front. The rear of the head is black and may develop some pruinescence with age. The pterostigma is black on the adult.

SIMILAR SPECIES: The Lyre-tipped Spreadwing (*L. unguiculatus*) is the most similar of the other spreadwings found in California, but it is typically less pruinose and more brightly colored above when mature (the dark dorsum of the abdomen is emerald green) and has extensive pale areas on the back of the head and pale tips to the pterostigma. Beware of crossed paraprocts on the male Common Spreadwing; these may superficially resemble the lyre-shaped paraprocts of Lyre-tippeds.

BEHAVIOR: Their behavior is similar to that of other spreadwings. They are often found ovipositing in tandem, preferably on green stems, in dense sedges of dried ponds and boggy meadows.

DISTRIBUTION: Although common and widespread throughout much of North America, this species is known in California primarily at elevations ranging from 760 to 2,400 m (2,500 to 8,000 ft) in the northern half of the state (Cascade-Sierran and Great Basin Provinces and the North Coast Ranges). It is also known from mountain ponds above Palo Alto in Santa Clara and San Mateo Counties and has been taken in the inner Coast Ranges in Solano and Contra Costa Counties. The only southern California record is from Los Angeles in the 1930s.

HABITAT: Like other spreadwings, this species breeds in fairly dense, low emergent vegetation bordering lakes, ponds, bogs, springs, and slow streams.

FLIGHT SEASON: The typical flight season ranges from July to September, but a few flight dates in June and even one in May have been recorded.

LYRE-TIPPED SPREADWING *Lestes unguiculatus*
Pls. 3, 4

LENGTH: 3 to 4 cm (1 to 1.5 in); **WING SPAN:** 3.5 to 4.5 cm (1.5 to 2 in.)

DESCRIPTION: The mature male is a rather bright spreadwing. The dark top of the thorax is iridescent bronze with pale green lateral stripes, and the top of the abdomen is metallic green. The paraprocts curve inward and then outward, forming a lyre-shaped structure from which the English name derives. The female and the immature male resemble the Common Spreadwing *(L. disjunctus)* but have a pterostigma with pale inner and outer tips. The rear of the head is extensively pale on the female.

SIMILAR SPECIES: The females of our other species of *Lestes* are entirely or mostly black on the rear of the head. The female Spotted Spreadwing *(L. congener)* may show pale areas on the rear of the head, but the pale dorsolateral stripes on the thorax are very narrow on the Spotted, whereas they are broad on the Lyre-tipped. The male Spotted has much shorter paraprocts. The male Common's crossed paraprocts may superficially look like "lyre tips," so careful examination in hand of the terminal appendages is necessary.

BEHAVIOR: They behave like other pond spreadwings, particularly Common Spreadwings, with which they are often found.

DISTRIBUTION: This is the least known species of spreadwing in the state. It has been found at just a few locations in California at elevations ranging from about 900 to 1500 m (3,000 to 5,000 ft) in the Cascade-Sierran and Great Basin Provinces from Placer County northward and at mountain ponds near Palo Alto in Santa Clara and San Mateo Counties, typically in small numbers with other, more common spreadwing species. It has a transcontinental distribution and is relatively common elsewhere.

HABITAT: It is found in freshwater marshes and around ponds and in lakes with extensive emergent vegetation.

FLIGHT SEASON: The Lyre-tipped Spreadwing flies for a brief period in summer, from mid-June through August.

POND DAMSELS
(*Coenagrionidae*)

Most of the world's damselflies, and the great majority (78 percent) of California's species, are in the huge pond damsel family (Coenagrionidae). They are mostly medium sized to small damselflies with stalked wings carried close together over the abdomen. Three large groups—dancers (*Argia*), American bluets (*Enallagma*), and forktails (*Ischnura*)—each with eight or 10 species found within our boundaries, constitute about three-fourths of the pond damsels in the state. In these groups and some other of our species, the predominant color pattern is black and blue or green on males and black and tan on females, but many species have two female color phases, one like males (andromorphic) and one not (gynomorphic) and usually duller. Many species inhabit ponds, hence the English family name, but some are more common along streams or rivers.

Dancers *(Argia)*

Although the name *Argia* signifies "laziness," dancers are actually quite active and alert damselflies. The genus contains well over 100 species, some still to be described, primarily within the American tropics and all in the Western Hemisphere. North American species total about 30, one-third of these reported from California. The diversity of species within the state decreases at higher latitudes, as might be expected of a group with primarily Neotropical affinities.

Males of most species have some blue or violet color on the thorax and abdomen, but two, the Sooty Dancer and the Powdered Dancer *(A. lugens* and *A. moesta),* are (as their names suggest) the color of soot and white ash, respectively. The vivid blues and purples dim considerably at low temperatures. Gynomorphic females and immature (or cold!) males are generally brown or gray where mature males are blue or violet, but andromorphic females exist as well in many species. The thorax pattern of many species involves a black stripe down the dorsal midline bordered on each side by a pale stripe, and below each pale stripe another black stripe, this often forked at its posterior end. Variations in this basic pattern are of use in identifying different species. Male dancers have distinctive pads, called tori, on the rear margin of abdominal segment 10, which are useful for identifying some look-alike species.

Most dancers, including all of our species, are associated with flowing waters, ranging in size from small spring runs to the larger rivers. Two or three species often coexist in suitable habitats. Pairs usually oviposit in tandem, sometimes in groups, the females using their ovipositors to deposit eggs into aquatic vegetation and waterlogged wood or other plant debris. Females or tandem pairs of some species may entirely submerge when ovipositing. The larvae are distinctive: short and flattened with stubby, broad gills.

The bluer species may be confused with American bluets *(Enallagma)*. Dancers show a preference to perch on the ground, rock walls, and other flat surfaces and usually carry their folded wings above the abdomen, whereas American bluets typically perch on vegetation with the wings folded alongside the abdomen, but there is overlap in these behaviors. Dancers exhibit a distinctive "wing-clapping" behavior. Often after returning to a perch, they partly spread their wings and then snap them back and together.

A close look at the spurs on the legs is a sure way to tell whether you have a dancer or American bluet (fig. 12). The tibial spurs of dancers are longer than the distance between their bases but are shorter, about equal to the distance between them, on bluets. Female dancers lack a vulvar spine, a feature present on female American bluets.

SOOTY DANCER   *Argia lugens*

Pl. 5

LENGTH: 4 to 5 cm (1.5 to 2 in.); **WING SPAN:** 5.5 to 6.5 cm (2 to 2.5 in.)

DESCRIPTION: Large and distinctively marked for a dancer, the mature male is dark and covered nearly all over with a sooty pruinescence. Even the wing membranes may look slightly smoky. At the bases of the middle abdominal segments are thin, pale rings of buff or tan, which are bright on the immature male and female but obscure on the mature male. Segments 8 through 10 are mostly dark above in both sexes. The female and the young male have an intricate pattern of tan and rusty brown (occasionally light blue) stripes on the black

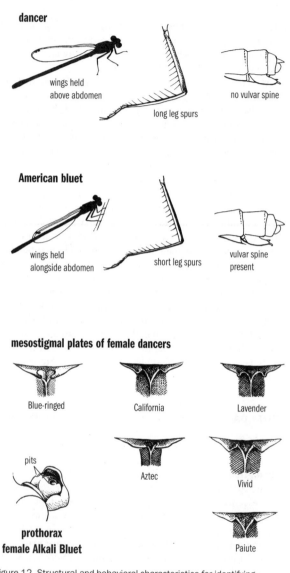

dancer

wings held above abdomen

long leg spurs

no vulvar spine

American bluet

wings held alongside abdomen

short leg spurs

vulvar spine present

mesostigmal plates of female dancers

Blue-ringed

California

Lavender

pits

Aztec

Vivid

prothorax female Alkali Bluet

Paiute

Figure 12. Structural and behavioral characteristics for identifying dancers and bluets. The flat top of each sketch indicates the hind margin of the prothorax, which obscures view of the anterior portion of the plate (see fig. 2).

thorax, with similar black and brown striping on the abdominal segments.

SIMILAR SPECIES: Other dancers found in California are differently marked, most noticeably in that abdominal segments 8 through 10 are pale above, and all except the Powdered Dancer (*A. moesta*) are smaller and daintier.

BEHAVIOR: Sooty Dancers are flighty, moving from rock to rock in the streambed in advance of an approaching observer. They land on driftwood or low branches over water but seldom in streamside vegetation. Females oviposit in growing stems in the stream. Males in tandem with ovipositing females do not perch stiffly erect as do other male dancers.

DISTRIBUTION: This is a southwestern species found primarily west of the Sierra Nevada in the California Province from Humboldt and Shasta Counties southward to the Mexican border, with a few reports east of the mountains in suitable habitat. Typically a species of the foothills, it is reported from sea level to about 1,200 m (4,000 ft).

HABITAT: The Sooty Dancer is a predictable denizen of gravel bars and the rock-strewn beds of streams and small rivers. It seems not inclined to wander far from the riparian corridor.

FLIGHT SEASON: This dancer has been found on the wing from April through October.

POWDERED DANCER   *Argia moesta*
Pl. 5
LENGTH: 4 cm (1.5 in.); **WING SPAN:** 4.5 to 6 cm (2 to 2.5 in.)
DESCRIPTION: This is a ghostly dancer, the pale thorax and somewhat darker abdomen heightening the illusion of a big-chested, long-tailed damselfly. The mature male looks as if it has been dipped in a fine powder or cold ash, with the head, thorax, and at least the basal and terminal (9 and 10) abdominal segments covered with a dense, white pruinescence. The middle abdominal segments are mostly black with variable light pruinescence and pale white basal rings. The thorax of the young male is broadly striped with brown and tan. The thorax of the female is pale tan or powder blue with a thin, dark midline on top; the abdomen is mostly tan or light blue with black lateral stripes, the middle segments having light basal rings, and segments 8 and 9 are pale above.

SIMILAR SPECIES: The Sooty Dancer (*A. lugens*) is as large, but darker and with the tip of the abdomen dark above. The female Emma's Dancer *(A. emma)* has the thorax mostly pale above, but it is smaller and its range in California does not overlap that of Powdered. The young Vivid Dancer *(A. vivida)* is chalky white but smaller and with well-defined, dark markings on the thorax and abdomen.

BEHAVIOR: Flying a few inches above the surface of light sand or alkali-encrusted shores, they are barely detectable, mostly by the movement of their shadows. They sit on the ground like other dancers but also perch with some frequency on streamside vegetation and the outer twigs of desert shrubs within a few feet of the ground. Nonbreeding individuals may forage some distance from water in weeds and desert scrub. Pairs may completely submerge when ovipositing in underwater vegetation, but they also oviposit at the surface on algal mats on rocks and drifted debris.

DISTRIBUTION: The Powdered Dancer is fairly common along the Colorado River and also at locations in and around the Coachella and Imperial Valleys, in San Bernardino, Riverside, and Imperial Counties. There is an old record from Los Angeles County (San Dimas). Most occupied sites are near, or even below, sea level.

HABITAT: This damselfly is found along or near the margins of spring runs, irrigation ditches, and rivers, typically in association with large rocks or debris.

FLIGHT SEASON: The known season is May through September, but it is probably more extensive.

BLUE-RINGED DANCER   *Argia sedula*
Pl. 6, Fig. 12

LENGTH: 3 to 3.5 cm (1 to 1.5 in.); **WING SPAN:** 4 to 4.5 cm (1.5 to 2 in.)

DESCRIPTION: This is a small and fairly dark dancer. The wings are often smoky or tinted amber. The adult male has a mostly black head with a pair of blue spots on top between the eyes. The thorax has a thick, black central stripe, two blue or violet blue lateral stripes bordered below by as thick or thicker black stripes with a small fork at the posterior ends, and lower sides blue, paler below. The abdomen is mostly black; the middle segments have blue basal rings, and segments 8 through 10 are blue above. The young male and the female are pale tan and brown where the mature male is blue and black. The darker brown markings on the thorax and the abdomen of the female are ill-defined.

SIMILAR SPECIES: The male is most similar to the male Paiute Dancer *(A. alberta)* and requires in-hand examination to distinguish. The Paiute Dancer lacks amber tint in its wings. In general, the black areas on the Blue-ringed Dancer's abdomen and thorax are more extensive than the pale areas; these are less extensive on the Paiute Dancer, but careful examination of differences in terminal appendages is necessary. Perhaps most easily seen with a hand lens in the field are differences in the tori: the hind margins curve inward to meet in a pointed V on the Blue-ringed Dancer but are aligned parallel to each other and separated by a fairly wide, U-shaped cleft on the Paiute Dancer. The female Paiute Dancer's dark dorsal markings on the thorax and abdomen are blacker and better defined than those of the female Blue-ringed Dancer.

BEHAVIOR: They behave similarly to other dancers, staying close to water and perching on the shore, floating debris, and low overhanging vegetation. Pairs oviposit in tandem on submerged and floating vegetation and woody, waterlogged debris.

DISTRIBUTION: The range in California is similar to that of the Powdered Dancer *(A. moesta),* but the Blue-ringed Dancer has also been collected as far north as Saratoga Springs and Death Valley National Monument, and eastward to interior San Diego County.

HABITAT: This species breeds along streams and rivers, but also at desert springs and oasis ponds.

FLIGHT SEASON: The Blue-ringed Dancer flies from March to November.

PAIUTE DANCER   *Argia alberta*
Pl. 6, Fig. 12
LENGTH: 3 cm (1 in.); **WING SPAN:** 4 to 4.25 cm (1.5 in.)
DESCRIPTION: This is a small, somewhat stocky damselfly. Its wing membranes are clear. The thorax has a fairly thick, black central stripe bordered by violet or blue stripes thicker than the black stripes below them, the latter forked posteriorly; the sides of the thorax are blue, paler below. The middle abdominal segments have pale basal rings but are otherwise black dorsally, the pale areas on the side separated by a narrow gap atop each segment. Abdominal segments 8 through 10 are light blue, with some small black marks on the sides. The pale areas on the young male

are light brown. The female is light brown and black, like the young male, but the pale areas on the sides of the middle abdominal segments are partly merged dorsally.

SIMILAR SPECIES: This species is most similar to the Blue-ringed Dancer *(A. sedula)* (see that species account for details on separating them). The black dorsum of the middle abdominal segments (except for narrow, pale rings basally) separates the male of this species from the male Aztec, California, Kiowa, and Lavender Dancers *(A. nahuana, A. agrioides, A. immunda,* and *A. hinei).* The female is not easily distinguished from females of these other species but lacks lobes on the posterior margins of the mesostigmal plates (fig. 12) that are present on all but the Kiowa Dancer. This feature is difficult to see, even with a high-powered hand lens.

BEHAVIOR: Paiute Dancers behave like other dancers. They are usually found in open desert areas, so typical perch sites are on the ground, or occasionally in low weeds over water.

DISTRIBUTION: The scientific name of this species is a bit misleading, as it has never been found in Canada. It was named by Clarence Kennedy for his father (Albert) and first collected in Inyo County, California. It has since been found widely distributed in the Great Basin, the western edge of the Great Plains, and the Southwest. In California, it is known from scattered sites in the Great Basin and Desert Provinces at elevations ranging from 1,500 m (5,000 ft) to below sea level.

HABITAT: This species is found at small streams and spring runs, including hot springs, the latter especially favored at higher latitudes (e.g., the Surprise Valley in Modoc County).

FLIGHT SEASON: The Paiute Dancer is on the wing from April through September.

KIOWA DANCER   *Argia immunda*
Pl. 6

LENGTH: 3.5 cm (1.5 in.); **WING SPAN:** 4 to 5 cm (1.5 to 2 in.)

DESCRIPTION: The mature male is blue or blue violet and black. The thorax pattern is similar to that of other dancers, mostly blue violet with a broad, black dorsal stripe and a thinner, black side stripe, forked at the posterior end. The abdominal pattern is distinctively banded, the middle segments having a pale basal ring and three dorsal bands (black, violet, and black) of roughly equal width. Segments 8 through 10 are blue above. The young male is light brown where the mature male is blue or violet. The female has a similar pattern, with pale areas blue (andromorphic form) or light brown (gynomorphic form). The dark stripe atop the female's thorax blends into pale areas on either side with a blurred, mottled effect.

SIMILAR SPECIES: Other male dancers in California with similar thoracic patterns (e.g., Aztec, California, and Paiute Dancers [*A. nahuana, A. agrioides,* and *A. alberta*]) lack the dorsal banding pattern on the abdomen typical of the Kiowa Dancer. The edges of the dark stripe atop the thorax of the female are less sharply demarcated than on the other species.

BEHAVIOR: They perch on streamside vegetation or on the ground and behave similarly to other dancers.

DISTRIBUTION: This dancer ranges from the southern Great Plains and American Southwest to Central America but has been found only once in California, at Resting Spring, Inyo County. It might occur at other springs in the Mojave and Colorado Deserts.

HABITAT: It is found along streams and spring runs.

FLIGHT SEASON: The California record is for late May, but the flight season elsewhere is from April to November.

LAVENDER DANCER   *Argia hinei*
Pl. 7, Fig. 12
LENGTH: 3 to 3.5 cm (1 to 1.5 in.); **WING SPAN:** 4 to 4.5 cm (1.5 to 2 in.)

DESCRIPTION: In life, the pale areas atop the thorax and on the middle abdominal segments of the mature male are a delicate lavender or pale violet color, unlike other dancers in its California range. On the thorax, these include violet dorsolateral stripes as wide or wider than the middorsal black stripe, and lavender violet sides below a relatively narrow, posteriorly forked, black side stripe. The lower sides and ventral surface of the thorax develop white pruinescence on the mature individual that contrasts with the violet dorsum. The middle abdominal segments are mostly lavender violet with relatively small, black markings; segments 8 through 10 are bright blue. The female and the young male are more or less light brown where the male is violet and blue.

SIMILAR SPECIES: The most similarly patterned species within its narrow range in the state are the California and Aztec Dancers

(*A. agrioides* and *A. nahuana*), males of which have pale areas that are blue or purplish blue, not lavender violet. Females of these species are similar in the field and require in-hand examination of mesostigmal plates (fig. 12) to confirm identification.

BEHAVIOR: Lavender Dancers perch on the ground, rocks, or low vegetation in the streambed. Tandem pairs oviposit in mats of aquatic vegetation and algae where narrow sheets of water trickle over rock slabs or gravel. Away from water, they forage in low weeds on roadsides and the edges of clearings on the surrounding slopes.

DISTRIBUTION: This is a species of the American Southwest and northern Mexico. In California, it is found only from Santa Barbara and Ventura Counties southward, on coastal slopes of the southern California mountain ranges.

HABITAT: The Lavender Dancer is found along small to medium-sized rocky streams and spring runs in the chaparral zone of foothill canyons.

FLIGHT SEASON: It is on the wing from May through October.

CALIFORNIA DANCER   *Argia agrioides*
Pl. 7, Fig. 12
LENGTH: 3 to 3.5 cm (1 to 1.5 in.); **WING SPAN:** 3.5 to 4.25 cm (1.5 in.)

DESCRIPTION: The mature male is typically bright blue and black on the head, thorax, and abdomen; the blue areas are sometimes a darker blue violet. The fairly broad, black stripe on the top of the thorax, and the thinner, black stripe on each side are usually forked (pattern quite variable) in the posterior half. The middle abdominal segments are mostly blue with a thin, black band at the rear end; segments 8 through 10 are blue above. The young male is brown where the older male is blue. The female may be gynomorphic (tan and black) or andromorphic (blue and black), with a similar pattern, but it has small, black stripes on the pale sides of the middle abdominal segments. See the "Similar Species" section for in-hand characteristics.

SIMILAR SPECIES: The Vivid Dancer, the Lavender Dancer, and especially the Aztec Dancer (*A. vivida, A. hinei,* and *A. nahuana*) are similar species, and all may occur together. The male Lavender Dancer can usually be distinguished by color because it lacks the blue tones of the California Dancer. The female Lavender Dancer has longer, black lateral stripes on the middle abdominal segments, confluent with a black band at the rear of each segment. The male Vivid Dancer has a black triangle in the blue side near the base of each middle abdominal segment, and both sexes usually lack the fork in the black stripe on the side of the thorax. The Aztec Dancer is identical in the field in all age and sex classes. In hand with magnification, each torus of the male California Dancer looks about as wide as the gap between them. Each torus of the male Aztec Dancer looks wider than the gap. The mesostigmal plates of the female California Dancer have small, knobby posterior lobes; the lobes on the Aztec Dancer are broad and flat edged (fig. 12).

BEHAVIOR: California Dancers behave similarly to other dancers. They perch on the ground, rocks near streams, and low stems over water. They are often found away from water, foraging in dirt paths and gravelly clearings in the stream or river channel, perched on the ground or short weed stems.

DISTRIBUTION: This species has a fairly restricted range centered, as the English name suggests, in California. It is fairly common and widely distributed west of the Cascade-Sierran Province and at desert springs and oases from Inyo County southward to Imperial County. It is also known from southern Oregon, Nevada, Arizona, and Baja California. It is primarily a species of valleys and foothills at elevations from below sea level near the Salton Sea to 1,500 m (5,000 ft) in Inyo County.

HABITAT: This species is found near a wide range of flowing-water habitats, from irrigation canals and rivers to small streams and spring runs. It occurs with the similar Aztec Dancer at a number of sites within the state, and broad overlap exists in the habitats occupied by the two species; however, the California Dancer seems to be more frequent at larger and deeper streams, whereas the Aztec Dancer is more frequent at tiny spring seeps and shallow, intermittent creeks.

FLIGHT SEASON: The adult of this species has been recorded on the wing from April to the first week of November.

AZTEC DANCER   *Argia nahuana*
Pl. 7, Fig. 12
LENGTH: 3 to 3.5 cm (1 to 1.5 in.); **WING SPAN:** 4 to 4.25 cm (1.5 in.)
DESCRIPTION: This species is very similar to the California Dancer *(A. agrioides)* (see the "Description" section of that species account for further details) and impossible to distinguish in the

field without in-hand examination with magnification. The two species were long confused, and many literature references to one actually refer to the other.

SIMILAR SPECIES: See the "Similar Species" section in the California Dancer species account for means of distinguishing the Aztec Dancer from the California Dancer and other similar dancers.

BEHAVIOR: Aztec Dancers behave similarly to other dancers.

DISTRIBUTION: Widely distributed in the western United States and Mexico, this species is also found in much of California in arid regions of the California, Great Basin, and Desert Provinces. It is apparently absent from the North Coast Ranges and the high Sierras. It is another valley and foothill species, and its highest known occupied elevations within the state are at 1,400 m (4,500 ft) in Modoc County.

HABITAT: It is found at spring runs, streams, and occasionally rivers in arid valleys, foothills, and desert country.

FLIGHT SEASON: This species has a rather long season, from March to November.

VIVID DANCER   *Argia vivida*
Pl. 8, Fig. 12

LENGTH: 3 to 3.75 cm (1 to 1.5 in); **WING SPAN:** 4 to 5 cm (1.5 to 2 in.)

DESCRIPTION: The English and scientific names aptly describe the brilliant blue or violet blue patches of color on the mature male. These areas of color are initially chalky white in the very young male, aging to lavender gray, then tan, and finally blue. Age and temperature can affect this color, which ranges from intense pure blue to deep violet blue to, when chilled, dull slate blue. Black marks on the mostly blue thorax include a broad dorsal stripe and a thin, black lateral stripe, the latter constricted to a hairline at its posterior end. Each middle abdominal segment has a thick blue band and a narrow black band, and a black triangle in the side of the blue band. Abdominal segments 8 through 10 are bright blue. The female is patterned like the male except the abdominal segments have more extensive pale areas. The pale colors of the female are highly variable, ranging from chalky white when young, to tan, lavender brown, light blue, or more rarely, as bright as the male.

SIMILAR SPECIES: The Vivid Dancer is best told from the Emma's Dancer *(A. emma)* by the thicker dorsal stripe on the thorax and the different color of the male. The California and Aztec Dancers

(A. agrioides and *A. nahuana)* are a bit smaller and more delicate, lack black triangular spots on the blue sides of their middle abdominal segments, and have a black stripe on the side of the thorax that is typically forked rather than constricted to a hairline. (Be cautious: rarely, the Vivid Dancer can have forked stripes, and one of the "branches" of the fork on the other two species is often obscure or absent.) The Kiowa and Lavender Dancers *(A. immunda* and *A. hinei)* have forked lateral stripes on the thorax and are differently colored and patterned. The white and black color of the young Vivid Dancer might lead to confusion with the Powdered Dancer *(A. moesta),* but the overall color patterns of the two species are very different.

BEHAVIOR: Vivid Dancers behave like other dancers. They perch in low vegetation along streams, as well as on the ground, logs, and rocks. Nonbreeding individuals are occasionally found foraging far afield in suburban yards and roadside weeds.

DISTRIBUTION: This dancer is widespread in the western United States, ranging from southern British Columbia and Alberta southward to Baja California. It has been recorded in every county in the state and is perhaps California's most ubiquitous damselfly. Most abundant in the interior valleys and foothills, it has been found from below sea level to over 2,100 m (7,000 ft) in the Sierra Nevada.

HABITAT: Its natural breeding sites are spring runs, seeps, and small foothill rivulets and streams. The Vivid Dancer has adapted to human activity, however, and also breeds in artificial drains in suburban neighborhoods and in irrigation ditches in rural areas. It is frequently found around ponds, lakes, and rivers, as well, but is probably breeding in the smaller streams feeding or draining these larger bodies of water.

FLIGHT SEASON: It has been recorded on the wing at all times of year in California, although it is very rarely seen in winter. The typical flight season in most low-lying areas is March through October.

EMMA'S DANCER   *Argia emma*
Pl. 8

LENGTH: 3.5 to 4 cm (1.5 in.); **WING SPAN:** 4.5 to 5 cm (2 in.)

DESCRIPTION: This is a medium-sized, long-legged dancer. The mature male is mostly violet and black above; the thorax is violet above and on the upper sides, with a thin, black midline and a thin, black side stripe that narrows to a hairline at its midpoint before widening again to a small, black spot near the base of the fore wing. The pale lower sides of the thorax develop a white pruinescence. The eyes are dark violet above. The middle abdominal segments are mostly violet above, segments 6 and 7 are more black than violet, and segments 8 through 10 are blue above. The young male is light brown where the mature male is violet. The female has a similar thoracic pattern, but the pale areas of the abdomen more extensive, with all dorsal pale areas a light olivaceous brown or with the top of the thorax and the anterior abdominal segments light blue.

SIMILAR SPECIES: Other dancers with which it occurs (usually the Vivid, Sooty, and California Dancers *[A. vivida, A. lugens,* and *A. agrioides]*) have distinctly different thoracic patterns. Sooty Dancers of both sexes are darker above and lack the pale terminal abdominal segments. The California and Vivid Dancers have a thick, black stripe on top of the thorax rather than a thin line; the males are blue or blue violet.

BEHAVIOR: These are active, wary dancers of open shorelines. They almost always perch on the ground or rocks, resorting to riparian vegetation for nightly roosts. When they land on a flat rock, they often spread their legs out and adopt a somewhat flattened, crouching posture from which they can quickly spring if approached, flying swiftly to another spot on the ground a few feet away. Females usually oviposit in tandem, either on the roots of aquatic vegetation while submerged or at the surface on waterlogged, algae-coated, woody debris.

DISTRIBUTION: This species has the most northerly distribution in the Pacific states of all dancers other than the Vivid Dancer, ranging from southern British Columbia and Montana to California, Nevada, Utah, and Colorado. In California, it occurs from the Oregon border southward to Monterey, Kern, and Inyo Counties at elevations ranging from sea level to 2,000 m (6,500 ft) in the Sierra Nevada.

HABITAT: Unlike our other dancers, the Emma's Dancer is often found along the shores of larger creeks, sloughs, and rivers. It is also found with other dancers on smaller streams. Mud banks littered with debris are visited, as well as rocky and sandy shorelines. It is even found along lakeshores, especially near stream mouths or outflows.

FLIGHT SEASON: This species flies from March through September.

Eurasian Bluets *(Coenagrion)*

As the name suggests, most species in this genus are found in the Old World. Only three of the 43 species of Eurasian bluets are found in North America, one of these reaching California. They resemble, and are probably closely related to, American bluets *(Enallagma)*.

TAIGA BLUET   *Coenagrion resolutum*

Pl. 9

LENGTH: 2.75 to 3 cm (1 in.); **WING SPAN:** 3.5 cm (1.5 in.)

DESCRIPTION: This is a small, inconspicuous damselfly. The male is mostly black above with pale blue (often decidedly turquoise blue) markings. The eyes are black above, turquoise below. The back of the head is mostly black with a large, blue spot behind each eye. The thorax is black above with two thin, blue dorsolateral stripes; these are often interrupted near the posterior end, thus resembling two exclamation marks. The sides of the thorax are turquoise blue blending into creamy yellow on the underside. The legs are striped black and turquoise. The abdominal pattern of the male is unique: segments 1 and 2 are mostly blue above with a black, U-shaped mark atop segment 2; segments 3 through 5 are half blue (basal half) and half black; segments 6 and 7 are mostly black above; segments 8 and 9 are mostly blue (sometimes with a thin, black line on the side of segment 8 and variable black markings on the rear half of segment 9); segment 10 is black above. The female's head and thorax pattern is similar to that of the male, the pale colors being either blue, turquoise, green, or tan. The female's abdomen is mostly black above, with narrow, interrupted rings at the bases of segments 4 through 7 and complete, narrow rings at the tips of segments 8 and 9. It has no vulvar spine.

SIMILAR SPECIES: Male American bluets within the range of the Taiga Bluet (primarily Boreal, Northern, and Tule Bluets *[Enallagma boreale, E. cyathigerum,* and *E. carunculatum]*) have different abdominal patterns and, especially, lack the black U atop segment 2. Female American bluets show more pale color at the bases of the abdominal segments above and have vulvar spines. The only other California damselfly with exclamation mark dorsal stripes is the larger Exclamation Damsel *(Zoniagrion exclamationis)* of lowland rivers and streams. The Swift Forktail *(Ischnura* *erratica)* female has a similar pattern but is larger, and the ranges do not overlap.

BEHAVIOR: Taiga Bluets fly low through dense vegetation and can be easily overlooked. After emerging they mature quickly, in about a week. Immature individuals may be found foraging within 100 m (300 ft) of the breeding site in low brush. Females, in tandem with males, oviposit in the stems of aquatic vegetation in shallow water. Taiga Bluets overwinter as late-instar larvae, which have been found in ice covered by snow. When appropriate temperatures are reached the following summer, all the larvae at a site emerge about the same time.

DISTRIBUTION: Of wide occurrence across Canada, the northern states, and the mountains of the West, this boreal species occurs at high elevations in the Cascade-Sierran Province from Shasta and Lassen Counties southward to El Dorado County. It has been found from 1,100 to 2,400 m (3,500 to 8,000 ft), but typically over 1,500 m (5,000 ft).

HABITAT: This species frequents dense, low, upright, emergent vegetation—sedges, grasses, horsetails—in bogs and along the margins of springs, ponds, lakes, and occasionally streams. Occupied sites often dry up, at least in part, in late summer and fall.

FLIGHT SEASON: The fairly short, summer season is from late May (more typically mid-June) to mid-August.

American Bluets *(Enallagma)*

American bluets are the commonly seen black-and-blue damselflies at marshes, ponds, and quiet backwaters of rivers and streams. Most frequent where emergent and floating vegetation intermingles with or borders patches of open water, their numbers may be so great in suitable habitat as to form shifting clouds of slender, blue lines darting low over the water's surface.

This is a predominantly Nearctic genus, most diverse in the eastern half of the continent. Of 35 species found north of Mexico, eight occur in California. American bluets are best represented in the state in the northeastern counties (Lassen and Modoc), where seven of the eight species (all but the tiny Double-striped Bluet *[E. basidens]* of the southern deserts) occur. In most parts of the state, at least two or three species are present, often at the same locality.

American bluets are easily confused with a number of other narrow-winged damselflies, especially the dancers *(Argia)* (see also the Taiga Bluet *[Coenagrion resolutum]*, Exclamation Damsel *[Zoniagrion exclamationis]*, and Swift Forktail *[Ischnura erratica]*). Dancers in general seem a bit more alert and "jumpy" and more often perch on the ground with their wings held above the abdomen like small sails, whereas bluets often perch on vegetation with their wings drooped alongside the abdomen. With the exception of the Double-striped Bluet, which if anything is more likely to be mistaken for a forktail because of its small size, all of our American bluets have a similar pattern on the pterothorax: a broad, black dorsal stripe (sometimes with a thin, pale midline) bordered on either side by a pale stripe, and below that a black stripe above pale sides, the pale areas being blue, gray, tan, or greenish depending on sex and age. The blue areas become darker at low temperatures. Dancers exhibit a greater variety of more complex color patterns, especially on the thorax. If in doubt, a good characteristic to look for in hand is the relative length of the tibial spurs (fig. 12). On bluets these are short, about equal to the distance between their bases. On dancers the spurs are longer than the distance between their bases.

NORTHERN BLUET   *Enallagma cyathigerum*

Pl. 9

LENGTH: 3 to 4 cm (1 to 1.5 in.); **WING SPAN:** 4 to 4.5 cm (1.5 to 2 in.)

DESCRIPTION: The mature male is among the brightest blue of our bluets, especially on the abdomen, the middle segments (3 through 5) of which are extensively blue above. Among California species in this genus, only the male Boreal and Familiar Bluets *(E. boreale* and *E. civile)* typically show as much blue on the top of the abdomen as does the Northern Bluet. The pale markings on the female may be tan, gray, greenish, or blue, and the top of abdominal segment 8 is partly to mostly pale above.

SIMILAR SPECIES: Other bluets (the River Bluet *[E. anna]*, Alkali Bluet *[E. clausum]*, and Familiar Bluet, but especially Boreal Bluet) are similar and usually not safely told apart in the field. The distinctive and relatively large cerci of the male River and Familiar Bluets may sometimes be visible at close range through binoculars, but in-hand examination of terminal appendages is the only sure way to identify the male of this species. Distinguishing the difference in the male appendages of Northern and

Boreal Bluets is particularly difficult, even with a high-powered hand lens. When viewed from the side, the superior appendages of both species appear short and rounded (shorter than the inferior appendages), but those of the Northern Bluet have a small, upturned hook projecting from the lower tip. The male Alkali Bluet has superior appendages that look wedge shaped in profile and project nearly as far back as the strongly hooked inferior appendages. The top of abdominal segment 3 is about half blue on the Alkali male, typically more so on the Northern. The female is similar to the female Boreal and Alkali Bluets in usually having the top of segment 8 pale, at least in part. It is not safely told from the female Boreal Bluet in the field.

BEHAVIOR: Similar to other bluets, breeding males patrol over water for females. Competition for perch sites on emergent vegetation is frequent where densities are high. The female oviposits in tandem into submerged vegetation, the attached male hovering in place above her. Nonbreeding individuals may wander some distance from water to forage in fields and meadows.

DISTRIBUTION: The Northern Bluet occupies not only most of North America but much of the Palearctic as well. In California it favors more humid regions and is scarce or absent in the drier parts of the Central Valley and the southern deserts; otherwise it is common and widespread, being recorded at elevations from sea level to over 3,000 m (10,000 ft).

HABITAT: Breeding sites run the gamut of aquatic habitats, from ponds, lakes, marshes, and bogs to springs, streams, and rivers.

FLIGHT SEASON: This species flies as early as mid-March in southern California, progressively later farther north and at higher elevations. It is on the wing through mid-October.

BOREAL BLUET   *Enallagma boreale*
Pl. 9

LENGTH: 3 to 4 cm (1 to 1.5 in.); **WING SPAN:** 4 to 4.5 cm (1.5 to 2 in.)

DESCRIPTION: This species is almost identical to the Northern Bluet *(E. cyathigerum)* in the field (see that species account for details). The female often shows a distinctive pattern on the top of abdominal segment 8: mostly pale with a black triangle, its base the rear edge of the segment and its apex toward the front. The Northern Bluet can have a similar pattern. The dorsum of abdominal segment 8 may be entirely pale on the female.

SIMILAR SPECIES: This species and the similar Northern Bluet are often found together. Inspection of the abdominal appendages of the male in hand with a high-powered lens may allow identification in some cases (see the discussion of differences in the Northern Bluet species account), but caution is necessary. The male Familiar Bluet *(E. civile)* has a similar color pattern, but its abdominal appendages are quite distinctive, the cerci being large and fin shaped. The female, which has all or most of abdominal segment 8 pale above, resembles the female Alkali Bluet *(E. clausum),* but females of all other bluet species except the Northern Bluet have this segment mostly black above. The female Northern and Boreal Bluets overlap considerably in color pattern and are not easily distinguishable in the field.

BEHAVIOR: Boreal Bluets behave similarly to Northern Bluets.

DISTRIBUTION: This species is widely distributed in the temperate regions of both the New World and the Old World. Within California, its distribution is much like that of the Northern Bluet. Common at mountain lakes in the north half of the state, it has been collected in the coastal ranges as far south as San Diego County. It is not known from the Desert Province or most of the Central Valley. Although found near sea level along the coast (e.g., the San Francisco Bay Area), it is usually found at elevations above 1,200 m (4,000 ft) in the interior.

HABITAT: This species is often found in numbers around mountain lakes with relatively modest emergent vegetation, but it is also found along creeks and in swampy areas. Boreal and Northern Bluets are frequently found at the same sites, although the numbers of one or the other usually predominate.

FLIGHT SEASON: The known span of the season in California—late April into early September—is rather short.

ALKALI BLUET   *Enallagma clausum*
Pl. 9, Fig. 12

LENGTH: 3.5 cm (1.5 in.); **WING SPAN:** 4.5 cm (2 in.)

DESCRIPTION: This is a robust bluet. The middle abdominal segments of the male are typically about half black and half blue.

The black stripe on the side of the thorax is often constricted posteriorly, but this is variable and other species can appear similar. The cerci of the male are wedge shaped in profile, whereas the paraprocts are strongly hooked upward. The female is distinctive among our bluets in having abdominal segment 8 completely pale and small pits (visible with a hand lens) on the upper lobes of the prothorax (fig. 12). The pale areas of the female are often distinctly olive green but also may be tan or pale blue.

SIMILAR SPECIES: The male Northern, Boreal, River, and Familiar Bluets *(E. cyathigerum, E. boreale, E. anna, and E. civile)* show as much or more blue on the tops of the middle abdominal segments. The Tule and Arroyo Bluets *(E. carunculatum and E. praevarum)* are typically less blue and more black on the middle segments and are in general smaller and more delicate. The abdominal appendages must be examined in hand to be sure of an identification. Of our other bluets, only some female Boreal and Northern Bluets may show a completely pale segment 8, but they usually have some black dorsal mark on that segment. The prothoracic pits are lacking in all our other species in the genus.

BEHAVIOR: These bluets have been characterized as occasionally behaving like dancers *(Argia),* warily flitting about on the alkali-encrusted edges of breeding ponds and lakes. Females oviposit in tandem in algal mats and other submerged vegetation. On windy days, which are common in their range, Alkali Bluets congregate with other odonates in the lee of bushes, away from water.

DISTRIBUTION: This large bluet inhabits the Great Basin Province southward to Inyo County at elevations in the range of 900 to 1,700 m (3,000 to 5,500 ft). It has been found as far west as Tule Lake along the northern border of the state.

HABITAT: The common name aptly indicates the habitat preference of this species in California. It is found around hot springs and in tule and cattail marshes and sedge meadows bordering alkaline and saline lakes, ponds, and occasionally creeks in sagebrush country.

FLIGHT SEASON: At the southern edge of its range, this bluet may fly as early as mid-May, but June through September is typical.

RIVER BLUET   *Enallagma anna*
PL. 10
LENGTH: 3 to 3.5 cm (1 to 1.5 in.); **WING SPAN:** 4 to 4.75 cm (1.5 to 2 in.)

DESCRIPTION: The male of this robust species has fairly broad, blue lateral stripes atop the thorax, and the middle abdominal segments are about half blue above. The male's cerci are striking and distinctive, each having a dorsal prong extending rearward considerably beyond the inferior appendages, and a pale tubercle on the inner surface. The female resembles other female bluets, especially the Familiar, Arroyo, and Tule Bluets (*E. civile, E. praevarum,* and *E. carunculatum*), with pale areas being either light blue or tan.

SIMILAR SPECIES: The Arroyo Bluet is quite similar but generally smaller and more slender. The cerci of the male Arroyo Bluet have less-pronounced dorsal prongs and lack the pale tubercles. Other bluets have dissimilar male abdominal appendages. The female is not easily told in the field from other female bluets, especially the Arroyo.

BEHAVIOR: River Bluets often alight on the ground and flit actively about along streams and the edges of dirt roads and trails like a dancer *(Argia).* Females, usually in tandem with males, may become completely submerged when ovipositing in emergent aquatic vegetation.

DISTRIBUTION: The River Bluet ranges through the highland valleys of interior western North America northward to southern Canada. In California, it is found primarily in the Great Basin Province southward to Inyo County, with a few reports from the upper Feather River drainage in Plumas and Sierra Counties, at elevations ranging from about 1,200 to 2,100 m (4,000 to 7,000 ft).

HABITAT: This species is somewhat unusual for a bluet, at least among our species, in being virtually confined to flowing waters. It is found along small streams and rivers, typically in valleys or on gentle slopes and fairly shallow areas with sandy or gravelly beds and grassy margins. It also occupies irrigation ditches and sloughs. Individuals wander to forage along the margins of nearby roads.

FLIGHT SEASON: Because of the relatively high elevations and, perhaps, the flowing-water habitats occupied by this species, its season is brief and late, from mid-June to mid-September.

ARROYO BLUET   *Enallagma praevarum*
Pl. 10
LENGTH: 3 to 3.5 cm (1 to 1.5 in.); **WING SPAN:** 3.5 to 4 cm (1.5 in.)
DESCRIPTION: This is a relatively delicate bluet, similar to the Tule Bluet (*E. carunculatum*) in overall appearance. Abdominal seg-

ment 3 is often more blue above than on the Tule Bluet. There is a tendency for the width of the dorsal blue band at the base of each of the middle abdominal segments (3 to 7) of the male to be about half that of the preceding segment, but this is variable. Reliable identification of the male requires examination of the cerci, two-pronged in side view, the top prong longest. The female is similar to other bluets, especially the Tule Bluet.

SIMILAR SPECIES: The Tule Bluet is most similar in overall size in pattern, and the two species may occur at the same sites. Differences in the cerci are noticeable in the male, but the female is not readily distinguished in the field, even in hand. See the "Similar Species" section of the River Bluet *(E. anna)* species account for ways in which that species differs from the Arroyo Bluet.

BEHAVIOR: Their behavior is similar to that of other bluets. They forage along the edges of riparian thickets but also away from water in adjacent uplands, including grasslands and oak savannas.

DISTRIBUTION: The Arroyo Bluet is found throughout most of southern California, along the coast northward to Sonoma and Napa Counties, in the Sierran foothills northward to Butte County, and in the sagebrush country along the northeastern border of the state. Recorded from sea level to 1,400 m (4,600 ft) in northeastern California, it is most frequent at intermediate elevations.

HABITAT: This southwestern species is typically found in relatively open, arid landscapes at ponds and the quiet backwaters of arroyos, washes, canyons, and river corridors.

FLIGHT SEASON: The Arroyo Bluet flies from March to October.

TULE BLUET   *Enallagma carunculatum*
Pl. 10

LENGTH: 2.75 to 3.75 cm (1 to 1.5 in.); **WING SPAN:** 3 to 4 cm (1 to 1.5 in.)

DESCRIPTION: This is a relatively small and fragile bluet. The male typically shows the least amount of blue dorsally on the middle abdominal segments (3 through 6) of any of our bluets. This pattern creates the illusion of a mostly dark, narrow-waisted abdomen flaring out to a pale blue tip. However, a male occasionally may show more than the usual amount of blue atop the middle segments and thus resemble other bluets, such as the Familiar and Arroyo Bluets *(E. civile* and *E. praevarum),* quite closely. The blue areas may have an aquamarine tint. Confident identification re-

quires examination of the male's terminal appendages with a
hand lens. The caruncle that gives this species its scientific name is
visible as a pale lobe extending from the tip of the slightly forked
superior appendage. The female is much like other bluet females.

SIMILAR SPECIES: Arroyo, Alkali *(E. clausum)*, and Familiar Bluets
all can look very similar and often coexist with this species. The
male's superior appendages may resemble a subdued form of the
Familiar Bluet male's larger, finlike ones, and the two species are
known to hybridize, so identification of some males may be
problematic. Superior terminal appendages of male Arroyo and
Alkali Bluets have distinctly different shapes and lack the pale
caruncles. Familiar and Arroyo females are not distinguishable in
the field from the Tule female.

BEHAVIOR: Tule Bluets behave like other bluets.

DISTRIBUTION: This is a very common species throughout much of
North America and most of California. It exhibits a considerable
elevational range, from sea level to at least 2,100 m (7,000 ft) in
the Sierra Nevada, but is especially common at lower to moderate
elevations throughout the state and frequently found with the
Familiar Bluet.

HABITAT: As implied by its English name, the Tule Bluet is common at marshy ponds and lakes, including those dominated by
tules *(Scirpus)* and other sedges. It also occurs along rivers and
streams with emergent vegetation and at temporary or disturbed
sites such as rice fields, irrigation ditches, and sewage ponds. It
seems to tolerate wide ranges of water quality and chemistry.

FLIGHT SEASON: This species is on the wing from February to November.

FAMILIAR BLUET   *Enallagma civile*
Pl. 10

LENGTH: 3 to 4 cm (1 to 1.5 in.); **WING SPAN:** 4 to 4.25 cm (1.5 in.)

DESCRIPTION: The mature male is a relatively large bluet with extensive amounts of blue on the middle abdominal segments. The
pale areas of the immature male (seen away from water) range
from pale blue to a lavender-tinged gray. Its distinctive cerci are
large and fin shaped, with a pale, triangular lobe along the trailing edge. They extend further back than the short paraprocts and
are easily viewed under modest magnification in the hand, sometimes even through binoculars at close range. The female is simi-

lar to many other female bluets and not easily distinguished. The females vary considerably in color, the pale areas on the thorax being either blue, light tan, greenish, or light gray. The pale areas of the middle abdominal segments often differ from the pale color on the thorax, being tan or greenish when the thorax is blue, or bluish when it is gray or tan.

SIMILAR SPECIES: The male Northern, Boreal, and Alkali Bluets *(E. cyathigerum, E. boreale,* and *E. clausum)* are similar in size and coloration. Tule and Arroyo Bluets *(E. carunculatum* and *E. praevarum)* often occur together with this species but are smaller, more delicate, and typically less blue on the abdomen. Some individuals of the latter two species, however, are bluer on the middle segments than normal, and hybrids between Familiar and Tule Bluets have been collected. The terminal appendages should be examined to be sure of an identification. The female is not safely told in the field from other bluets such as the Tule or Arroyo.

BEHAVIOR: Much like other bluets, breeding males congregate over water and at shoreline vegetation. Oviposition is in tandem. Females and immature males forage in fields, meadows, and suburban gardens, often far from water. They forage by hovergleaning from low vegetation, perching briefly between bouts to handle and chew prey. Like other bluets, they may steal prey from spider webs, taking the spiders themselves at times. The tables can turn, however, with the web catching the bluet! This species is often encountered in urban settings, even gleaning prey from the windshields of parked cars.

DISTRIBUTION: Widespread in North America, this is a familiar species in much of California as well. It apparently increased in numbers and spread north within the state in the last century as a result of human activity. Most abundant at lower elevations within the California Province and in the southwest corner of the state, it also occurs in the Great Basin Province, along the coast to the Oregon border, and at a few locations in the Sierra Nevada below 1,800 m (6,000 ft).

HABITAT: This species frequents artificial bodies of water, such as stock and wastewater ponds, reservoirs, irrigation ditches, and rice fields. It is often one of the most abundant odonates at such sites. It also occurs at freshwater marshes and along rivers and creeks, especially where heavily affected by human activity.

FLIGHT SEASON: This species flies as early as March and continues into early December. Numbers seem to peak at many sites in late summer, and it remains common well into fall.

DOUBLE-STRIPED BLUET   *Enallagma basidens*
Pl. 10

LENGTH: 2 to 2.75 cm (1 in.); **WING SPAN:** 2.5 to 3 cm (1 in.)

DESCRIPTION: This is a tiny bluet that is easily overlooked in dense, low vegetation near water. Once spotted, however, it is readily identified by its small size and the presence of two pale lateral stripes on either side of the thorax (the lower stripe thin and difficult to see without magnification). There is also a thin, pale stripe down the center of the thorax. These features apply to the female, which usually has the pale areas tan rather than blue, as well as the male. The female has abdominal segment 10 entirely pale.

SIMILAR SPECIES: This elfin species is much smaller than our other bluets and readily distinguished from them by the two pale lateral stripes. In size it resembles the Citrine Forktail *(Ischnura hastata),* with which it coexists, but the latter is markedly different in color (the male shows no blue and has a yellow-tipped tail; the female is either extensively orange or pruinose and shows at most only one pale stripe on each side of the thorax above).

BEHAVIOR: These are unobtrusive bluets that stick close to vegetation near water and appear to be weak fliers. This species' widespread dispersal in recent decades is thus hard to explain, unless perhaps it has been assisted by human activity such as the creation of large-scale water projects.

DISTRIBUTION: This species is known in California only from the vicinity of the Colorado River along the border with Arizona, from Riverside and Imperial Counties. Elevations in its range in California are within a few hundred feet above sea level. This species underwent a considerable range expansion in North America in the last century, spreading northeast- and westward from the southwestern border states, and it is apparently a fairly recent colonizer of California.

HABITAT: The Double-striped Bluet occurs in marshy river backwaters, ponds, sloughs, and ditches. It seems to tolerate relatively stagnant and polluted waters.

FLIGHT SEASON: This species has been seen in California from July through September but no doubt flies earlier in summer, too.

Exclamation Damsel *(Zoniagrion)*
The lone species in this genus, a California endemic, is similar to both the American bluets (*Enallagma*) and the forktails (*Ischnura*).

EXCLAMATION DAMSEL   *Zoniagrion exclamationis*

Pl. 11

LENGTH: 3 to 3.5 cm (1 to 1.5 in.); **WING SPAN:** 4 to 4.5 cm (1.5 to 2 in.)

DESCRIPTION: The typical male and many females somewhat resemble American bluets in being bright blue and black but differ in having each dorsal stripe on the pterothorax split into a line and a spot—forming the "exclamation marks"—and in having most of the abdomen black above except for blue markings on segments 1 and 2 and a patch of blue at the tip (on segments 7 through 9 on males and segments 7 and 8 of females). Many females and the occasional male have the dorsal thoracic stripes constricted rather than interrupted, shaped something like little baseball bats. The older female may be darker, the pale blue areas becoming slate colored or purplish, and is difficult to distinguish from older individuals of some other species such as bluets and forktails. The female has a pronounced vulvar spine.

SIMILAR SPECIES: The only other California damselflies with interrupted dorsal stripes are the Taiga Bluet *(Coenagrion resolutum)* (a smaller species with a very different abdominal pattern found at high mountain bogs) and some andromorphic female Pacific Forktails *(Ischnura cervula)*. The latter are small and delicate, with only the top of abdominal segment 8 pale. An Exclamation Damsel with complete dorsal stripes perhaps most closely resembles the male and andromorphic female Swift Forktail (*I. erratica*) in overall pattern. The dorsal stripes of the Swift Forktail are even sided, not shaped like baseball bats as in this species. The male forktail has prominent dorsal spikes (the "forktail") at the rear edge of segment 10 and very different abdominal appendages. The andromorphic female Swift Forktail has a weak or absent vulvar spine and a blue dorsal spot on abdominal segments 7 through 9 (segment 9 is all dark above on the female Exclamation Damsel).

BEHAVIOR: The foraging behavior of Exclamation Damsels is similar to that of bluets. Individuals search through vegetation in riparian clearings a few feet above the ground for prey and are not prone to wander far from water. They often occur in partly shady situations, flitting from one sunlit patch to the next. Females oviposit alone, above and below the water line, in emergent vegetation (e.g., bur-reed *[Sparganium]*) bordering quiet, muddy backwaters.

DISTRIBUTION: The known range includes the coastal zone, Central Valley, and Sierran foothills from Mendocino and Shasta Counties southward to Santa Cruz and Merced Counties, at elevations from around 600 m (2,000 ft) to sea level.

HABITAT: The species frequents the vicinity of quiet pools and backwaters of rivers and streams and, occasionally, ponds bordered by willows *(Salix)*, alders *(Alnus)*, and such. It is particularly associated with sun-dappled, sylvan glades of blackberries *(Rubus)* and herbaceous vegetation such as poison-hemlock *(Conium maculatum)*, mugwort *(Artemisia douglasiana)*, and Queen Anne's lace *(Daucus carota)*.

FLIGHT SEASON: The early and relatively brief season lasts from March into the first week of August.

Forktails *(Ischnura)*

This is a rather diverse genus of damselflies containing many species and found nearly worldwide. Of the 14 species in the contiguous United States and Canada, eight occur in California. In many parts of the state, one or more forktail species are often the most abundant damselflies in dense stands of low emergent vegetation such as sedges and grasses.

Forktails vary somewhat in appearance and size. Perhaps the most distinctive structural feature is the forked tail, which is actually a two-pronged projection at the upper hind tip of abdominal segment 10 on males, although this is weakly developed in some species. In addition, males and some females of all our species but one (the Citrine Forktail *[I. hastata]*) have the top of the abdomen mostly black with a blue patch atop some of the terminal segments (usually 8 and 9). Males of most species (except for the Black-fronted and San Francisco Forktails *[I. denticollis* and *I. gemina]*, which comprise the subgenus *Celaenura*) have a pair of thin, pale stripes or four pale spots (the Pacific Forktail *[I. cervula]*) atop the thorax. Females in some species are particularly variable in appearance, occurring in gynomorphic and andromorphic color phases as well as changing in color with age, usually from bright, contrasty colors to dull pruinescence.

SWIFT FORKTAIL   *Ischnura erratica*

Pl. 11

LENGTH: 3 to 3.5 cm (1 to 1.5 in.); **WING SPAN:** 3.5 to 4 cm (1.5 in.)

DESCRIPTION: This damselfly is large and robust for a forktail. The head is mostly black with a green stripe on the face. The lower half of the eye is green, and it has blue spots on top between the eyes. The thorax is black above with two rather broad, blue lateral stripes; the sides are blue or aqua, shading to yellow green below. The abdomen is mostly iridescent black above, yellow green to yellow orange below, with blue bands atop segments 1 and 2, narrow yellow rings on bases of segments 3 through 6, and a large blue patch atop segments 8 and 9 extending onto the base of segment 10 and often onto rear half of segment 7 as well. The prominent, upward-projecting forktail at the tip of segment 10 and the long, rearward-projecting paraprocts of the male are often visible through binoculars at close range. The male's pterostigma in the fore wing is bicolored, black toward the base of wing and tawny toward the tip. The andromorphic female is like the male, but blue areas may be greener. The gynomorphic female has pale areas on the head, and the thorax is green, with abdominal segments black above except for narrow, green rings near the base. The pale areas on the head and the thorax of the immature female are a dull orange. All females have a tan pterostigma and lack a prominent vulvar spine.

SIMILAR SPECIES: Other forktails in its range are noticeably smaller and more delicate and have different color patterns. The male and andromorphic female most closely resemble Exclamation Damsel *(Zoniagrion exclamationis)* individuals that show complete blue stripes on the thorax. The stripes atop the thorax of the Exclamation Damsel, if not interrupted, are constricted at the posterior end (so they look like little baseball bats). The male Exclamation Damsel lacks the pronounced forktail and long, projecting inferior appendages of the male Swift Forktail, and the female has a pronounced vulvar spine and blue color atop segment 9. Greenish female bluets somewhat resemble the gynomorphic female Swift Forktail, but they have a distinct vulvar spine and their middle abdominal segments have pale sides, with wedge-shaped extensions that converge to a point atop each segment at its base.

BEHAVIOR: Swift Forktails behave more like bluets or dancers than other forktails. They are more likely to perch in open situations and on horizontal surfaces (lily pads, the ground) and when flushed often fly high into a tree or bush, or out over open water. They may be found foraging along weedy margins of roadsides and trails away from water. Females oviposit, unattended by males, into floating vegetation or plant stems under the surface.

DISTRIBUTION: This species has a restricted range along the Pacific Coast from British Columbia southward to northern California. It occurs in the Northern Coastal Province southward to Santa Clara and San Mateo Counties. Most known locations are close to sea level, a few to about 300 m (1,000 ft).

HABITAT: The Swift Forktail occupies small, marshy ponds and lakes with clear, shallow water and emergent vegetation, including ponds in river floodplains. These typically have wooded shores with a lush herbaceous understory.

FLIGHT SEASON: This is an early flyer, with typical dates ranging from late March into June, rarely found as late as September.

WESTERN FORKTAIL   *Ischnura perparva*
Pl. 11

LENGTH: 2.5 to 3 cm (1 in.); **WING SPAN:** 2.5 to 3 cm (1 in.)

DESCRIPTION: The male is a small, mostly black-and-green damselfly. The top of the head is black with a green spot beside each eye. The eyes are black above and green below. The thorax is black above with two narrow, green to aqua stripes and green (occasionally aqua) on the sides, shading to yellow below. The abdomen is mostly black above and yellow below, with thin, green basal rings on the middle segments and a blue patch covering segments 8 and 9. The pterostigma of the fore wing is black. The female is relatively stout, especially the abdomen, compared to other small forktails. The young gynomorphic female has a head and thorax pattern something like that of the male, but the pale areas are orange and the abdomen is black above except that segments 1 through 3 (sometimes 4) and 9 and 10 are orange. The mature gynomorphic female has pale areas on the thorax that are a dull green but are scarcely visible beneath a gray pruinescence covering all of the upper surfaces. The pruinescence on the middle abdominal segments is pale gray. The pterostigma is smoky gray. An extremely rare teneral andromorphic female has been collected on a few occasions. It has pale areas on the thorax that are tan, and abdominal segments 8 through 10 are pale blue above.

SIMILAR SPECIES: The Swift Forktail *(I. erratica)* has a somewhat similar pattern but is much larger, has a restricted range, and has blue dorsal stripes on the thorax of the male. The mature female Pacific Forktail *(I. cervula)* is darker, nearly black, and not as stocky.

BEHAVIOR: These rather tame forktails make short, low flights above the dense beds of sedges and grasses in which they breed. They occasionally wander a short distance from water to forage in grassy meadows and fields. Females oviposit, typically alone, by submerging the abdomen and laying eggs on the stems of emergent vegetation just below the surface.

DISTRIBUTION: The Western Forktail is common and widespread, especially in northern California, but is found statewide, except in the southeastern Desert Province, at elevations ranging from sea level to at least 2,400 m (8,000 ft) in the Sierra Nevada. This species inhabits much of the western United States.

HABITAT: It is found in low emergent vegetation of springs, bogs, ponds, and lakes, as well as pools and backwaters of streams and rivers. It may forage over nearby grassy areas.

FLIGHT SEASON: This species has a long season over its extensive range, from March to November, with the earliest and latest dates at lower elevations.

RAMBUR'S FORKTAIL   *Ischnura ramburii*
Pl. 13

LENGTH: 2.75 to 3.5 cm (1 to 1.5 in.); **WING SPAN:** 3 to 4 cm (1 to 1.5 in.)

DESCRIPTION: The male is a good-sized forktail. The head is mostly black with some green color on the face and with fairly small blue or green spots on the back of the head near the eyes. The eyes below shade from aqua to yellow green. The top of the thorax is black with a pair of thin, green, golden green, or aqua stripes; the sides are green or aqua fading to yellow green below. The slender abdomen shades from green (on the basal segments) to tawny yellow on the sides and undersurfaces of segments 3 through 6, with a more or less parallel-sided black stripe above from segment 2 to segment 6. Most of segment 7 is black with blue patches low on the sides next to segment 8, which is nearly all blue. Segments 9 and 10 are usually black above and blue below and on the sides, but segment 9 may be all or partly blue above. The lower legs (tibiae and tarsi) are mostly black. The andromorphic female is similar to the male, but green areas may be bluer. The gynomorphic female's thorax is mostly pale, with a black dorsal stripe and the basal abdominal segments pale. The rest of the abdomen is black above except for occasional pale spots atop segments 9 and 10. The pale areas on the gynomorphic female are often bright, glowing orange but may be tan or olive green with age. The female has a pronounced vulvar spine on segment 8.

SIMILAR SPECIES: The only similar species in its California range is the Desert Forktail *(I. barberi)*, the male of which always has abdominal segment 9 blue above, the middle abdominal segments with dart-shaped black patches on top, and broader, yellow green stripes atop the thorax. This pattern on the middle abdominal segments also distinguishes the female Desert Forktail from the female Rambur's Forktail. In addition, the gynomorphic Desert Forktail female is paler, not as bright orange as many Rambur's Forktail females, and has the top of abdominal segment 8 pale.

BEHAVIOR: These big forktails will go after large prey, including other damselflies and even tenerals of their own species. Oviposition, as in many other forktails, is done by the female alone, by curling her abdomen under floating vegetation.

DISTRIBUTION: This damselfly is little known in the state and only recently discovered here. All records have been in the Desert Province along the Colorado River, around the Salton Sea, and in the Imperial Valley. This is one of the most common and widespread species of damselfly in the Western Hemisphere, however, found from the northeastern United States southward to Chile, and its eventual discovery elsewhere in southern California would not be surprising.

HABITAT: This species is found at desert springs, marshy ponds and lakes, and backwater lagoons along the Colorado River.

FLIGHT SEASON: The few records for California to date are from May to December, but this forktail is found flying the year around elsewhere in its range.

DESERT FORKTAIL   *Ischnura barberi*
Pl. 13
LENGTH: 2.75 to 3.5 cm (1 to 1.5 in.); **WING SPAN:** 3 to 4 cm (1 to 1.5 in.)

DESCRIPTION: This species is similar to Rambur's Forktail *(I. ramburii)*, but its areas of pale coloration are somewhat more extensive and pastel hued. The top of the head and eyes are black, with fairly large, pale blue spots between the eyes connected, or nearly so, by a thin, pale line. The eyes below are blue to aqua. The top of the thorax is black with two yellow green to aqua lateral stripes nearly half as wide as the black dorsal stripe between them. The sides of the thorax are pale pastel blue or aqua. The pattern on the middle abdominal segments is distinctive: black above, pale orange yellow below and on the sides, with the black area atop each segment pointed at the base and constricted to a hairline subterminally, then widening again to form a terminal spot. The effect produced, as viewed from above, is of a row of small, black darts. Segments 8 and 9 are blue above and below, and segment 10 is black above and blue on the sides. The lower legs are mostly pale. The andromorphic female is much like the male, with small, black spots on segment 9 above. The gynomorphic female differs from the andromorphic female in having a mostly pale thorax with a variable, usually narrow, dark dorsal stripe; her pale areas are tan or tawny orange with some green tints on the abdomen. The female has a well-developed vulvar spine.

SIMILAR SPECIES: The Desert Forktail occurs at the Salton Sea and along the Colorado River with the similar Rambur's Forktail. The male Rambur's Forktail often, but not always, has the top of abdominal segment 9 black, and the gynomorphic female is usually brighter orange, with the top of segment 8 black. It is best told from that species by the shape of the black areas atop the middle abdominal segments (no subterminal constriction on Rambur's Forktail) and the narrower, green stripes on top of the thorax of the male Rambur's Forktail.

BEHAVIOR: This little-known species appears to have habits similar to those of other forktails.

DISTRIBUTION: This species is of widespread but local occurrence in the Desert Province northward to southern Inyo County. Along the coast, it is found in San Diego and Orange Counties. Occupied sites range from below sea level around the Salton Sea to about 1,100 m (3,500 ft) on the high deserts. The range of the species is limited to the southern United States west of the Mississippi River.

HABITAT: Desert springs, marshy ponds, lakes, and lagoons are habitats occupied by this species. This forktail is often found at, but not restricted to, alkaline or brackish waters, such as those around the Salton Sea and in coastal lagoons, with dense borders of grasses, sedges, and other emergent vegetation.

FLIGHT SEASON: This species has an extended season, from March to November.

PACIFIC FORKTAIL   *Ischnura cervula*
Pl. 12

LENGTH: 2.5 to 3 cm (1 in.); **WING SPAN:** 2.5 to 4 cm (1 to 1.5 in.)

DESCRIPTION: The four pale spots in each corner of the black, rectangular top of the thorax readily identify the male. These spots are almost colorless in the first few hours after emergence but quickly become pale blue. The sides of the thorax and the eyes below the black upper surface are pale blue or aqua, shading to yellow green below. The top of the head, including the eyes, is black with blue spots between the eyes on top. The abdomen is black above, green to yellow below, with the top of segments 8 and 9 blue. The pterostigma is black. Females show considerable variation. Most commonly seen are the immature female, which has a mostly pale thorax (pale tan or lavender gray to nearly white) with a black dorsal stripe of variable width and a thin, black side stripe, salmon pink spots on the back of the head between the eyes, and a black abdomen with the top of segment 8 pale blue; and the fully mature gynomorphic female, which is mostly pruinescent black or slate gray above, yellow green below, and olive green on the sides of the thorax. In between these stages, the pale areas on the head and thorax may be blue or lavender. The rarer andromorphic female has four blue spots atop the thorax, like the male, the front two often teardrop shaped, and only abdominal segment 8 is blue above. The pterostigma color is nearly white on the young female, smoky or tan on the older female. The female's prothorax has a distinctive central lobe along the rear margin flanked by a pair of incurved pencils (pointed tufts) of hairs, the latter resembling thin, sharp spines under 8× to 10× magnification.

SIMILAR SPECIES: No other damselfly in our area has four pale dots on the top of the thorax, but these are sometimes hard to see in the field, in which case Black-fronted and San Francisco Forktails (*I. denticollis* and *I. gemina*) can look similar. The dark, pruinose female is difficult to tell from some other species that become mostly dark above with age or at cold temperatures, so identify with caution. No other species have the projecting pencils of hairs present on the back rim of the prothorax.

BEHAVIOR: Of all our forktails, these seem most prone to wander some distance from water to forage in weedy fields and suburban yards. In dense emergent vegetation, they dart low into cover if disturbed. Females usually oviposit unattended in sedge stems or while perched on floating vegetation, curling the abdomen underwater to deposit eggs on the undersides of leaves.

DISTRIBUTION: Found throughout much of western North America from southern Canada to northern Mexico, this forktail is recorded from nearly every county in California. It is most abundant at lower elevations but occurs, at least locally, in the Cascade-Sierran Province up to 2,200 m (7,300 ft).

HABITAT: This species occupies many aquatic habitats, including small roadside ditches, springs, marshes, boggy ponds, lakes, slow streams, and riparian backwaters. Breeding sites often have the water surface coated with algae or duckweed *(Lemna)* or filled with floating plants. Nonbreeding individuals forage in grassy or weedy areas away from water.

FLIGHT SEASON: One of the first odonates seen in numbers in spring, this species has been found flying year-round in the state.

BLACK-FRONTED FORKTAIL   *Ischnura denticollis*
Pl. 12

LENGTH: 2 to 2.5 cm (1 in.); **WING SPAN:** 2.5 to 3 cm (1 in.)

DESCRIPTION: This is the smallest damselfly found in most of California. In most of the state, the male is easily identified by the pattern on the thorax: solid iridescent black above, without stripes or spots, and green or aqua on the sides. The head is black above with two small blue or green spots between the eyes, which are black above and green below. The abdomen is black above, aqua on the lower sides of segments 1 through 3 and 8 through 10, and yellow on the lower sides of segments 4 through 7, and has blue patches atop segments 8 and 9. There is a projecting hook at the lower tip of each paraproct. The rare andromorphic female resembles the male. The gynomorphic female has a pattern similar to that of the male, but with only the central stripe on the thorax black, bordered by broad, pale lateral stripes that are separated from the pale sides by a thin, black stripe. The colors of the young female are a soft coral pink on the pale dorsal surfaces of the head and thorax (also coral pink legs), shading to white on the sides and below. The abdomen is black above, and small pale spots, usually green, may or may not be present atop abdominal segments 8 and 9. With age, the female becomes pruinose and darker above. The female's prothorax has two dorsal hornlike projections, visible in hand with a good lens.

SIMILAR SPECIES: The San Francisco Forktail *(I. gemina)* is nearly identical and hybridizes with the Black-fronted Forktail in some parts of its limited range, further complicating the picture. In-hand identification is required where both occur (San Francisco Bay Area). The female San Francisco Forktail lacks protuberances on the prothorax, whereas the male lacks the projecting spine at the lower end of the inferior appendage. Old, dark female Citrine or Pacific forktails *(I. hastata* and *I. cervula)* can resemble an old female of this species, but both lack the protuberances on the prothorax.

BEHAVIOR: These are rather weak-flying, delicate forktails that tend to stay low in emergent vegetation in shallow water. They do not wander very far. This species and the closely related San Francisco Forktail are unique among forktails in that the females typically oviposit in tandem with males. While both are perched on a sedge stem or grass blade, she dips her abdomen in the water and places a few eggs in the vegetation just below the water line. The pair then moves a short distance and repeats the procedure.

DISTRIBUTION: The Black-fronted Forktail is widespread and fairly common in southern California, northward in the California Province to Shasta and Sonoma Counties, and in the Great Basin Province northward to the Oregon border. It is typically found in arid habitats below 1,500 m (5,000 ft), but is unrecorded along the North Coast and in the higher mountains. It ranges from the American Southwest to Guatemala.

HABITAT: Dense, low sedges and beds of grass near springs, small ponds, streams, and the still backwater lagoons of rivers are typical haunts. This forktail is often found at hot springs and is not inclined to wander too far from water.

FLIGHT SEASON: This species has been seen on the wing much of the year, from March to November.

SAN FRANCISCO FORKTAIL — *Ischnura gemina*
Pl. 12

LENGTH: 2.5 to 2.75 cm (1 in.); **WING SPAN:** 2.5 to 3.25 cm (1 to 1.5 in.)

DESCRIPTION: The name *gemina* refers to the fact that this species is a virtual twin of the more widespread Black-fronted Forktail *(I. denticollis).* In-hand characteristics include the shape of the male abdominal appendages viewed from the side, the cerci having a blunt, dorsal lobe projecting rearward, the paraprocts lacking projecting hooks. The female lacks protuberances on the prothorax. The side of the thorax on the male is generally more blue or aqua in color than on the Black-fronted Forktail, but this varies in both species.

SIMILAR SPECIES: See the Similar Species section for the Black-fronted Forktail for comparisons. The male Pacific Forktail (*I. cervula*) has four pale spots on the top of its thorax, but these are difficult to see at times. The dark female Pacific Forktail has pencils of hairs on the rear margin of the prothorax.

BEHAVIOR: San Francisco Forktails behave much like Black-fronted Forktails. They wander only short distances from water to forage in weedy fields.

DISTRIBUTION: Endemic to California, this forktail is known only from the San Francisco Bay Area. It has been recorded northward to Sonoma County (Bodega Bay), southward to Monterey County (the Salinas River), and as far east as Contra Costa and Alameda Counties.

HABITAT: This species frequents small, marshy ponds and ditches with emergent and floating aquatic vegetation. It is not often found far from water.

FLIGHT SEASON: It has been seen flying as early as March and as late as November.

CITRINE FORKTAIL   *Ischnura hastata*
Pl. 13

LENGTH: 2 to 2.5 cm (1 in.); **WING SPAN:** 2 to 3 cm (1 in.)

DESCRIPTION: This is the smallest North American damselfly. The tiny male is unmistakable, primarily because the abdomen is mostly sulfur yellow, with reduced black markings on the middle segments, segments 8 through 10 nearly all yellow, and only segments 1, 2, and 7 mostly black above. The forktail projecting from atop segment 10 is long, slender, and yellow. Another unique feature on the male is the free-floating pink or tawny pterostigma in the fore wing, which is not connected to the wing edge. This contrasts with a more normally positioned, smaller, black pterostigma in the hind wing. The head (including the eyes) and thorax of the male are mostly black above with small, green dots atop the head between the eyes and two thin, yellow green lines atop the thorax. The lower eyes and sides of the thorax are yellow green, and the legs are mostly pale yellow. The female is always gynomorphic; colors vary with age. The young female is orange on the face, with the top of the head black, except for large, orange spots rearward, and has a broad, black stripe atop the thorax above orange sides that fade to white or yellow green below. Abdominal segments 1 through 4 (sometimes 5) and 9 and 10 are mostly orange; segments 6 through 8 are extensively black above. Pale colors become reduced in extent and brown with age, and eventually the entire upper surfaces are covered with gray pruinescence. The pterostigma of the female is tan.

SIMILAR SPECIES: The Citrine Forktail is found along the Colorado River with the similarly sized but differently colored Double-striped Bluet *(Enallagma basidens),* which has two pale dorsolateral stripes on the thorax, the male with blue tops to abdominal segments 8 and 9, the female with segment 10 all pale tan. Other forktails in range are larger or have blue atop segments 8 and 9 on the male and some pale markings atop these segments on the female. The dark female Pacific and Black-fronted Forktails *(I. cervula* and *I. denticollis)* are typically a bit bulkier but overlap in size. In-hand examination of the prothorax with an $8\times$ to $10\times$ lens may be necessary to eliminate these species. The bright orange immature female somewhat resembles the Desert Firetail *(Telebasis salva),* but the latter species lacks black marking on the abdomen.

BEHAVIOR: These frail forktails stick very close to saltgrass and dense sedge beds bordering ponds and lagoons. They are easily overlooked because of their small size. They are apparently easily dispersed by winds if carried aloft, so occasionally they appear far from water. Females typically oviposit alone on aquatic vegetation.

DISTRIBUTION: This species is found in the Desert Province, primarily in the vicinity of the Salton Sea and the Colorado River, eastward to Los Angeles County and northward into southern San Bernardino County. Occupied sites are at or near sea level. This species is widely distributed from southeastern Canada to South America.

HABITAT: It prefers dense, low emergent vegetation around ponds, springs, irrigation ditches, and riparian backwaters.

FLIGHT SEASON: Recorded flight dates are from April through September.

Sprites *(Nehalennia)*

The six species in this genus are widely distributed from the taiga to the tropics: one across Eurasia; one in Central America, the

Antilles, and South America; and four in North America. The one species found in California is also found across the northern United States and Canada.

SEDGE SPRITE   *Nehalennia irene*
Pl. 14

LENGTH: 2.5 to 3 cm (1 in.); **WING SPAN:** 2.75 to 3.25 cm (1 to 1.5 in.)

DESCRIPTION: This is a true sprite, so small and slim bodied it is almost impossible to see in dense grass. The eyes are blue, the top of the head and thorax are metallic blue green to emerald green, and the lower sides of the thorax are creamy white. The abdomen is metallic emerald green above and white to creamy yellow below, with patches of gray pruinescence atop parts of segments 8 and 9 and nearly all of segment 10. The legs are striped black and pale yellow.

SIMILAR SPECIES: The metallic green thorax separates the Sedge Sprite from other odonates of comparable size in California. The Emerald Spreadwing *(Lestes dryas)* has a vaguely similar color pattern, but it is a bulky giant compared to this elfin species and, of course, carries its wings spread.

BEHAVIOR: Sedge Sprites are rather unwary and deliberate. Flying short distances through dense grasses and sedges, often stopping to perch, they are nearly invisible. Pairs oviposit in tandem on stems of emergent vegetation below the water line. They have occasionally been found along the shoreline of mountain lakes near breeding sites.

DISTRIBUTION: This sprite is known from only a handful of bogs and swamps (Willow Lake, Plumas County; Cooper Swamp, Lassen County) in the Lassen Peak region at elevations around 1,800 m (6,000 ft).

HABITAT: In California, it inhabits dense beds of grasses and sedges in shallow bogs and swampy meadows.

FLIGHT SEASON: Adults have been found from June through August.

Firetails *(Telebasis)*

These slim damselflies would be easily overlooked if not for their bright red color. The 40 species in this genus are primarily found in the tropics, from Mexico to South America, but two range into the southern United States. One species occurs in California.

DESERT FIRETAIL   *Telebasis salva*
Pl. 14

LENGTH: 2.5 to 3 cm (1 in.); **WING SPAN:** 2.5 to 3 cm (1 in.)

DESCRIPTION: This is an elegant little damselfly. The male has a relatively long, slender abdomen that is scarlet red above, paler below, without conspicuous dark marks. The eyes are also scarlet red. The thorax is dull red above, fading to yellow on the lower sides and undersurface, with two parallel, black dorsal bars separated by a thin, pale midline, and a smaller black mark on the side. The leg color is a pale tan. The female pattern is similar to that of the male but duller, tawny brown.

SIMILAR SPECIES: Young females of some forktail species, especially the Citrine Forktail *(Ischnura hastata),* can be mostly orange, but they have some dark markings on the abdomen. The Western Red Damsel *(Amphiagrion abbreviatum)* occupies different habitats for the most part, has a much shorter abdomen with noticeable dark markings on the terminal segments, and has dark legs.

BEHAVIOR: When flying low over mats of bright green vegetation, the red males are a striking sight. Desert Firetails often tend to stick tight in dense, marshy cover, though, making them surprisingly easy to miss. Females in tandem oviposit in mats of algae or other floating aquatic vegetation.

DISTRIBUTION: This little damselfly has an extensive range, from northern California southward to Venezuela. Here at the northern end of its range, it occupies relatively arid lowlands and foothill country west of the Sierra Nevada from Mendocino, Glenn, and Shasta Counties southward to Mexico, and the southern deserts as far north as Inyo County on the eastern slope of the mountains. It has been recorded from below sea level (near the Salton Sea) to about 1,200 m (4,000 ft).

HABITAT: Springs, ponds, lakes, and the still backwaters and stagnant pools in stream beds and river beds may all be occupied by the Desert Firetail. A key habitat requirement appears to be dense, floating mats of green algae, duckweed *(Lemna),* or similar aquatic vegetation.

FLIGHT SEASON: The Desert Firetail flies from April through October.

Red Damsels *(Amphiagrion)*

Small, stout damselflies in the genus *Amphiagrion* are found throughout much of North America. Populations in the West and the East exhibit minor but consistent differences and are commonly thought to represent two separate species (with intermediate populations in midcontinent perhaps representing a third species). In any event, all California populations are of the western form.

WESTERN RED DAMSEL   *Amphiagrion abbreviatum*
Pl. 14

LENGTH: 2.5 to 3 cm (1 in.); **WING SPAN:** 3 to 4 cm (1 to 1.5 in.)

DESCRIPTION: This stubby damselfly has a short abdomen that barely extends beyond the tips of the folded wings. The head and thorax of the male are dark above, mostly black; the abdomen is red with large, black dorsolateral patches on segments 7 through 10. The legs are mostly black. The female is usually tawny red or tan throughout with similar proportions, and rarely darkly pruinose on the entire upper surface. In hand, a distinctive mark visible on both sexes is a hair-covered bump on the undersurface of the thorax behind the legs.

SIMILAR SPECIES: Our only other small red damselfly, the Desert Firetail *(Telebasis salva),* is slender and delicate, with about a third of its total abdomen length extending beyond the tip of the folded wings; the male's thorax is mostly red and lacks the underside thoracic bump. The Desert Firetail is a species of arid lowlands; its range rarely overlaps that of the Western Red Damsel, but they are found together at a few spots in the Sierra Nevada foothills.

BEHAVIOR: Western Red Damsels fly low in dense emergent vegetation and are rarely seen far from water. Often, they seem to be either extremely common or very scarce at a particular locality, suggesting perhaps highly synchronized emergence and a relatively short breeding season at any one site.

DISTRIBUTION: This species avoids the more arid, low-lying areas of the state but is otherwise widely distributed, especially in mountainous areas. It is found down to sea level along the northwest California coast and in the Coast Ranges from Humboldt County to Marin County, the Great Basin from Modoc County to Inyo County, the Cascade-Sierran Province to nearly 3,000 m (10,000 ft), and mountain ranges west of the deserts in southern California from Ventura and Kern Counties southward to San Diego County.

HABITAT: This species is found at small ponds, springs (including hot ones), seeps, bogs, and the borders of streams and lakes. Shallow water with dense beds of low, emergent vegetation—sedges, rushes, grasses—is typically occupied.

FLIGHT SEASON: This species flies from April to September, with the peak season in June and July.

PETALTAILS
(Petaluridae)

Petaltails are primitive dragonflies, throwbacks to forms that were the dominant odonates in Jurassic times and no doubt circled about the heads of dinosaurs. Only 10 species exist in the world today, most of these in the Southern Hemisphere. The genus *Tanypteryx* contains two species, one in Japan and the other in the Pacific Northwest, including northern California.

BLACK PETALTAIL *Tanypteryx hageni*
Pl. 20, Figs. 8, 11
LENGTH: 5.5 cm (2 in.); **WING SPAN:** 7.5 cm (3 in.)

DESCRIPTION: The adult is a large, black-and-yellow dragonfly whose dark brown or blackish compound eyes don't come in contact. The yellow areas on the top of the middle abdominal segments have a distinctively ornate U shape. The shape of the pterostigma is also unique: inordinately long and narrow. The English name derives from the male's abdominal appendages, which are broad and shaped something like the petals of a flower. The female has a stout ovipositor. The immature petaltail is not as bright as the adult; the black and bright yellow areas are replaced by slate gray and creamy yellow.

SIMILAR SPECIES: The Pacific Spiketail *(Cordulegaster dorsalis)*, often found at the same bogs as the Black Petaltail, has yellow stripes on the thorax instead of spots, is larger, has bright blue eyes that just meet atop the head, and behaves very differently. Clubtails also have separated compound eyes, and some are black and yellow (e.g., Grappletail *[Octogomphus specularis]*), but these are somewhat smaller and have very different color patterns, a pterostigma that is not strongly attenuate, and a club-shaped abdomen in the male.

BEHAVIOR: Males at breeding sites are rather unwary and easily approached. They defend small territories at suitable sites in the bog, awaiting the arrival of females. Females oviposit directly onto the mossy substrate. This species exhibits a strong tendency to perch on flat surfaces, both horizontal (rocks, fallen logs, paved roads) and vertical (trees, posts, and even people). The larvae occupy fairly deep burrows, from the mouths of which they ambush prey.

DISTRIBUTION: The Black Petaltail is found at scattered bogs in the Northern Coastal and Sierran-Cascade Provinces southward to Sonoma, Napa, and Mariposa Counties. Elevations occupied range from within about 100 m (a few hundred feet) of sea level at some sites near the coast to over 2,100 m (7,000 ft).

HABITAT: This species inhabits bogs and seeps, where the stocky, primitive larva dwells in a burrow in the mossy muck. The adult is seldom far from the bog, occasionally wandering to nearby clearings in the forest.

FLIGHT SEASON: This is a summer species, on the wing from mid-May through mid-August.

DARNERS (Aeshnidae)

The mostly large, impressive dragonflies in this family have a characteristic profile on the wing—big head (mostly eyes!), bulky thorax, and long, slender abdomen, the latter supposedly resembling a darning needle and thus giving rise to the common name. Other characteristic features include the eyes in broad contact atop the head and a well-developed ovipositor on the female. The wings generally lack dark markings (other than the pterostigma) but are tinted with amber or brown in some species. The larvae are long-bodied, active predators that stalk all manner of prey including small tadpoles and fish, as well as their own kind (fig. 7).

Green Darners (Anax)

These are big, awe-inspiring dragonflies. One of our two species is the largest dragonfly in North America, and the other is one of the most common and widespread. They are characterized by large size, a plain, green thorax, and a long abdomen marked with blue on males (and some females). The intensity of blue on the abdomen is temperature dependent, fading to a dull purple or green when cold. When hanging up to roost in green vegetation, they can be quite difficult to spot. They are quick and agile hawkers, foraging, often at dusk, from near ground level to high overhead.

COMMON GREEN DARNER *Anax junius*
Pl. 15

LENGTH: 6.5 to 8 cm (2.5 to 3 in.); **WING SPAN:** 9 to 10 cm (3.5 to 4 in.)

DESCRIPTION: This is a large, handsome dragonfly with a robust green thorax and, on the male, a mostly sky blue abdomen with a broad, dark brown dorsal stripe. The face is yellow green with a striking bull's-eye pattern (a black spot ringed with blue and yellow) atop the frons. The large compound eyes are olive or brown, more opalescent on the young individual. The male's cerci are slender, simple, and spine tipped. The female and the immature male are similar to the mature male, but the abdomen color is often dull green, reddish purple, or gray brown. The female has two stubby horns along the rear edge of the occiput between the eyes (an in-hand characteristic). The wings are often amber tinted.

SIMILAR SPECIES: The Giant Darner *(A. walsinghami)* is larger, more elongate, and of local occurrence. Other darners have a striped thorax. The male Western Pondhawk *(Erythemis collocata)* has a green thorax and blue abdomen during the transition to mature adult coloration but is smaller, with a green face and different proportions (a shorter, stubbier abdomen).

BEHAVIOR: Males patrol over water along constantly changing routes. They aggressively pursue potential mates to the point of slamming them into vegetation or even attempting to break up tandem pairs. They are unique among our darners in that oviposition often occurs in tandem, but females also oviposit unattended. Females oviposit in emergent and submerged plant stems and waterlogged vegetative debris. This species often forages in swarms over rivers, lakes, meadows, even roads and towns. Apparently there are both resident and migratory populations. Residents emerge in late spring and midsummer, breed, and produce larvae that overwinter. Migrants emerge in the Tropics and arrive in early spring to reproduce, their offspring emerging in late summer and fall. Immature migrants swarm in mountain meadows in August and September and are seen moving in large numbers downslope along rivers and streams, headed south to breed and renew the cycle.

DISTRIBUTION: This species is found statewide, from sea level to at least 2,300 m (7,500 ft), and throughout the contiguous United States, as well as northward into southern Canada and Alaska and southward to Honduras and the West Indies. It also occurs in Hawaii and Bermuda and has strayed to East Asia, Great Britain, and islands in the South Pacific Ocean.

HABITAT: The Common Green Darner breeds in a wide range of still waters, including ponds, lakes, marshes, sloughs, and backwater lagoons of creeks and rivers. Nonbreeding individuals, especially during migration, may be found foraging almost anywhere.

FLIGHT SEASON: This species has been seen on the wing year-round but is most widespread and abundant in August and September.

GIANT DARNER   *Anax walsinghami*
Pl. 15

LENGTH: Males, 10 to 12 cm (4 to 4.5 in.); females, 9 to 10 cm (3.5 to 4 in.); **WING SPAN:** 11 to 12 cm (4.5 in.)

DESCRIPTION: This is North America's largest dragonfly. The head, thorax, and wings are only a little larger than those of the Common Green Darner *(A. junius),* but the abdomen is exceptionally long and slender, longer than 6.5 cm (2.5 in.) and up to 9.0 cm (3.5 in.) on the male. It is carried with a noticeable arc by the male in flight. The face is yellow green, and the frons is topped with a bull's-eye mark like that of the Common Green Darner. The eyes are dark olive with blue tints. The male has a green thorax, with some blue toward the back end, and a long, dark brown abdomen with segments 2 and 3 blue and a broken blue stripe down each side that becomes increasingly fragmented rearward. The male's cerci are broad and paddle shaped. The female is similar, but has a shorter (but still very long) abdomen, the sides of which are marked with tan, green, or blue.

SIMILAR SPECIES: The only dragonfly remotely similar is the Common Green Darner, which is considerably smaller, especially in terms of abdomen length (less than 6.0 cm, or about 2 in.), has narrower, simpler cerci on the male, and has a pair of short horns on the rear of the occiput on the female.

BEHAVIOR: Males fly long beats down stream courses, returning to the same spot only after a long interval. They are extremely wary. When females arrive at the stream, they are quickly captured and carried off in high, rapid flight.

DISTRIBUTION: Records of this species in the state come primarily from the California Province, in foothill canyons in Kern, Ventura, and Los Angeles Counties and southward and in the inner Coast Ranges as far north as Colusa County, at elevations below 900 m (3,000 ft). Records from Inyo and Shasta Counties may represent isolated populations or wanderers. It ranges southward to Central America.

HABITAT: The Giant Darner occurs along permanent, spring-fed streams in arid country. It is not frequently seen away from water.

FLIGHT SEASON: This species has been reported flying from April to September in California.

Mosaic Darners *(Aeshna)*

This is a large genus of big, brown dragonflies that are striped and spotted with blue, green, and yellow. Nearly 80 species are found worldwide, most in the Northern Hemisphere, and a quarter of these in North America. Diversity is greatest at higher latitudes and elevations and at lakes, ponds, and boggy meadows in conifer forests, where six to nine species may be found at the same site in some parts of the continent. In California, four or five species can be found in the Cascade-Sierran Province in late summer and fall.

All species in the genus share certain characteristics. They have fairly pale faces that contrast with large, dark or bright blue eyes, and a striking black T-spot atop the frons. There is a distinctive strip of black and sky blue in the upper surface of each compound eye. The thorax is robust, the abdomen long and slender. The brown thorax has pale stripes of blue, green, white, or yellow on the sides and (more noticeably in males) the front. The abdomen has a mosaic pattern of blue spots on males and blue, green, and/or yellow on females. The blue areas on some species change from bright at higher temperatures to dull when cooler. They are strong fliers that actively hawk for prey and roost by hanging from vegetation or perching on the sides of tree trunks. Females oviposit unattended by males, inserting eggs into aquatic vegetation or woody, waterlogged debris.

The seven species known to occur in California can be grouped as follows:

Blue-eyed Darner *(A. multicolor):* One widespread species, the most common mosaic darner at lower elevations. The male has bright blue eyes and distinctive, forked cerci.

California Darner *(A. californica):* A small, widespread species that flies earlier than other mosaic darners. It shares some features with the Blue-eyed Darner (e.g., tubercle on underside of segment 1). The male has simple cerci like the species in the next group.

Typical mosaic darners: Two species (Variable and Canada Darners *[A. interrupta* and *A. canadensis]*) of mountain lakes. The males have relatively simple, straight-edged cerci.

Paddle-tailed darners: Three species (Shadow, Paddle-tailed, and Walker's Darners *[A. umbrosa, A. palmata,* and *A. walkeri]*). The males have cerci that are twisted and somewhat paddle shaped in profile, with a noticeable spine at the lower tip.

Field identification of most mosaic darners to species is a considerable challenge and usually impossible in the case of flying individuals. Perched males studied in detail at close range may often be identified using a combination of the characteristics illustrated here, but the group is best learned by netting individuals and examining them in hand. This is easier said than done, however, because they are wary and fast. Most females can be confidently identified only in the hand.

BLUE-EYED DARNER — *Aeshna multicolor*
Pls. 16, 19

LENGTH: 6.25 to 7 cm (2.5 to 3 in.); **WING SPAN:** 8.5 to 10 cm (3.5 to 4 in.)

DESCRIPTION: The male is our most striking and distinctive mosaic darner, appearing very blue in the field. The face is pale blue without a black line between frons and clypeus, and the eyes are bright blue. The stem of the T-spot widens slightly at the base, and the crossbar is fairly thin. The stripes on the sides of the thorax are straight sided, fairly broad, and blue. The black abdomen has a mosaic pattern of blue and rust spots. The male's cerci are distinctively forked near the tip (when viewed from the side). The female is similar in pattern but variable in color, often quite multicolored in blue, purple, green, and yellow. The female often has the wings washed with amber between the nodus and the pterostigma. Both sexes have a small tubercle on the underside of abdominal segment 1. The pale areas on the abdomen change in color with temperature.

SIMILAR SPECIES: The California Darner *(A. californica)* is most similar to the male Blue-eyed in color, but it is smaller and has duller blue eyes, simple-type cerci on the male, and a black stripe across the face in both sexes. In profile, the California Darner has a somewhat thicker, shorter abdomen but a less robust thorax with thinner pale stripes. Other female mosaic darners (except the California Darner) are distinguished in hand by the lack of a small bump on the underside of abdominal segment 1.

BEHAVIOR: Males patrol a few feet above water, with frequent hovering and changes of direction, along shoreline, or over open patches of water in marshes. Individuals, especially females and immature males, forage widely, often a considerable distance from breeding sites along roads, in meadows and yards, and even on open mountain tops and offshore islands. They may forage in small groups or in larger swarms with other species.

DISTRIBUTION: This is a common species, especially at lower elevations. It is found statewide, including offshore islands, from sea level to over 2,100 m (7,000 ft). The Blue-eyed Darner is widespread and common in much of western North America.

HABITAT: It breeds in marshes, ponds, lakes, and still stretches and backwaters of streams and rivers, tolerating a wide range of water chemistry and pollution levels.

FLIGHT SEASON: The Blue-eyed Darner has been seen on the wing as early as mid-March and as late as November.

CALIFORNIA DARNER   *Aeshna californica*
Pl. 16, 19

LENGTH: 5.5 to 6 cm (2 to 2.5 in.); **WING SPAN:** 8 to 8.5 cm (3 to 3.5 in.)

DESCRIPTION: Our smallest darner, it has a more evenly tapered profile than other mosaic darners, without as much contrast between the stocky thorax and narrow waist. The face is pale blue gray with a black stripe between the frons and clypeus and a thick T-spot stem widening at the base. The eyes are blue gray, browner below. The male has a brown thorax with small, pale spots on the front and narrow, pale blue lateral stripes. The abdomen is dark brown with large, pale blue dorsal spots (duller when cold) on all segments. The male's cerci are simple in shape. The pterostigma is brown. The female may be colored like the male or have yellow green stripes on the thorax and green abdominal spots above. Both sexes have a tubercle on the underside of segment 1.

SIMILAR SPECIES: The Blue-eyed Darner *(A. multicolor)* male has bright blue eyes and broad, blue stripes on the thorax. Both sexes of the Blue-eyed lack the black stripe across the face, have wider stripes on the thorax, and are slightly larger (but some approach the California Darner in size). The Variable Darner *(A. interrupta)* is larger, darker overall, and generally flies at higher elevations later in the year. The male and female are distinguished in hand from all our other darners, except for the Blue-eyed, by the tubercle under segment 1.

BEHAVIOR: Their foraging behavior is much like that of other darners. Patrolling males fly low circuits along the creek or shoreline.

DISTRIBUTION: This darner is common throughout most of California at elevations ranging from sea level to about 1,800 m (6,000 ft). It is apparently absent from the Desert Province and the higher mountain regions. Despite its name, this species occurs throughout much of western North America, from British Columbia to Baja California.

HABITAT: This species breeds in marshy ponds, lakes, streams, and river backwaters lined with emergent vegetation. Occupied habitats include alkaline marsh ponds and hot springs pools east of the Sierras. It is found foraging away from water along roads and in clearings, including suburban yards.

FLIGHT SEASON: It flies early for a darner, from March (at lower elevations) into August.

VARIABLE DARNER   *Aeshna interrupta*
Pls. 17, 19, Fig. 7
LENGTH: 6.5 to 7 cm (2.5 to 3 in.); **WING SPAN:** 9 cm (3.5 in.)
DESCRIPTION: This species looks darker than other mosaic darners in the field. The face is pale olive with a black stripe. The stem of the T-spot is fairly thick with a wide base, and the eyes are dark blue green. The thorax is brown with, at most, thin, short, pale blue stripes on the front (but these are often missing) and two thin, pale blue lines on each side, one or both of the lateral stripes often interrupted in the middle to form two spots or short dashes. The abdomen is black with blue spots on the top of all segments. The male's cerci are of the simple type. The pterostigma is brown. The female has a similar pattern but is quite variable in color, the pale areas on the thorax and abdomen being blue, green, or yellow or some combination thereof. The yellow-marked female has brown eyes and amber-tinted wings.

SIMILAR SPECIES: No other mosaic darner in California has interrupted stripes on the side of the thorax. If the thorax stripes are complete, it most resembles and often flies at the same sites with the Paddle-tailed Darner *(A. palmata)*. The male of the latter species has thicker, yellow green side stripes on the thorax and brighter blue abdominal markings; it looks much brighter overall. The paddle-shaped cerci of the Paddle-tailed are visible at close range on a perched male. The female Paddle-tailed can be very similar and difficult to tell apart, even in hand, but it always has complete, usually thicker side stripes on the thorax. The

Shadow Darner *(A. umbrosa)* lacks the dark stripe on the face, has a mostly or entirely black top of segment 10, and, in hand, exhibits tan on the back of the head and blue spots on the underside of the abdomen. The California Darner *(A. californica)* is smaller and paler. The Canada Darner *(A. canadensis)* lacks the black face stripe and has distinctively shaped stripes on the thorax and blue spots below on the abdomen.

BEHAVIOR: Mature males patrol the borders of mountain lakes and sedge-lined ponds in much the same manner as other mosaic darners. Females oviposit in stems of emergent vegetation, typically just below the surface of the water. Both males and females wander some distance from water to feed, occasionally in swarms along roads and clearings. They forage from near ground level to a considerable height, showing great agility on the wing. When roosting, they seem more likely than other mosaic darners to perch low on the side of a bush, in low weeds, or even on rocks or the ground. They also perch on tree trunks.

DISTRIBUTION: The Variable Darner is widespread and fairly common in the Cascade-Sierran Province south to Tulare County at elevations ranging from 1,400 to 3,400 m (4,500 to 11,000 ft). There are also scattered records for the Great Basin southward to Mono County and the North Coast Ranges southward to Glenn County. This a common species across Canada, the northern states, and the mountains and high plains of the West.

HABITAT: This species inhabits mountain lakes, boggy ponds, and stream pools in sedge meadows.

FLIGHT SEASON: The flight period is from June (rarely in May) into early October.

CANADA DARNER   *Aeshna canadensis*

Pls. 17, 19

LENGTH: 6 to 7 cm (2.5 to 3 in.); **WING SPAN:** 8.5 to 9 cm (3.5 in.)

DESCRIPTION: This darner has a pale blue green face without a strong, black stripe between the frons and clypeus, gray green eyes, and black on the back of the head. The anterior lateral stripe on the thorax resembles a small shoe; it is distinctly notched in the middle and rounded at the lower end, with a small posterior projection at the top. This stripe is often bluer above and greener below. There are paired, blue spots on the undersides of the middle abdominal segments. The male's cerci are of the simple type.

SIMILAR SPECIES: Other, similar mosaic darners in our area have an anterior lateral stripe on the thorax that is either straight (California, Shadow, and Paddle-tailed Darners [A. californica, A. umbrosa, and A. palmata]) or reduced to a thin, complete or broken line (Variable Darner [A. interrupta]). The Shadow Darner is our only other species with blue spots on the underside of the abdomen, but the male has paddle-like cerci, and in both sexes the back of the head is tan.

BEHAVIOR: They behave much like other darners. Males fly low routes along the lakeshore or along channels in the marsh or bog, in search of females. They also forage in nearby open forest, taking meadowhawks and similarly sized prey.

DISTRIBUTION: Only recently discovered in the state at scattered locations in the Cascade-Sierran Province (Siskiyou, Tehama, Plumas, and Lassen Counties), this darner probably has been overlooked and is more widely distributed. It is common across southern Canada and the northern United States. Recorded elevations range from around 1,100 to 1,800 m (3,500 to 5,500 ft).

HABITAT: The Canada Darner has been found at marshy ponds and mountain lakes, typically with floating or emergent vegetation.

FLIGHT SEASON: The few California records range from mid-August to early October.

SHADOW DARNER   *Aeshna umbrosa*
Pls. 18, 19
LENGTH: 6.5 to 7.75 cm (2.5 to 3 in.); **WING SPAN:** 8.5 to 10 cm (3.5 to 4 in.)

DESCRIPTION: This is a more or less typical mosaic darner when seen in the field, but with a distinctive combination of characteristics visible in hand. The face is a dull, gray green without a dark cross stripe, and the T-spot has a short, thin stem and a straight-edged crossbar. The eyes are blue gray to brown. The back of the head is tan on both sexes. The thorax is brown with distinct, green frontal stripes on the male and straight, lateral stripes that are yellow to green below and often blue above. The blue spots on top of the abdomen are relatively small and typically pale blue (darker at colder temperatures). Usually, segment 10 lacks blue spots above, and segments 4 through 6 have paired blue (or lavender gray in some females) spots on the undersides. The male's cerci are paddle shaped with fairly long spine at the lower tip, similar to the cerci of the Paddle-tailed Darner *(A. palmata).*

The pterostigma is brown. The female typically has yellow green abdominal spots above, but these may be blue like those of the male. The female's wings are tinted brown.

SIMILAR SPECIES: Paddle-tailed and Walker's *(A. walkeri)* Darners are quite similar, especially the females. Differences observable in the field are few, but on perched individuals allowing close inspection, Paddle-tailed and Variable *(A. interrupta)* Darners show a black stripe across the face and blue spots atop segment 10. Walker's Darner has whitish lateral stripes on the thorax and black pterostigma. In hand, all our other mosaic darners have black on the back of the head, and the Canada Darner *(A. canadensis)* is the only other one with pale spots on the underside of the abdomen.

BEHAVIOR: The English name refers to the patrolling habits of males, which carefully search for ovipositing females within a foot of the water in the shadowy nooks and crannies along the shores of tree-shaded streams. Females oviposit in mud or moss banks and in soggy logs and other wooden debris along the shore.

DISTRIBUTION: Widespread across much of North America, this darner is common in the Cascade-Sierran Province southward to Tulare County and in the Northern Coastal Province southward to Santa Cruz County. It is also present in the foothill country of the California Province around the San Francisco Bay Area and the Sacramento Valley. Recorded elevations range from sea level to 2,300 m (7,500 ft).

HABITAT: This species is found in a wide range of aquatic habitats in wooded or forested country. Most often found along meandering streams with pools and riffles, it also occurs at marshy ponds, bogs, and mountain lakes. It can be found away from water along roads and in clearings.

FLIGHT SEASON: The peak flight period is from late July into early November, but there are a few records for May and June. An amazing mid-February record from Marin County may represent an unusually timed emergence in a drought year.

PADDLE-TAILED DARNER   *Aeshna palmata*

Pl.18, 19

LENGTH: 6.5 to 7.5 cm (2.5 to 3 in.); **WING SPAN:** 8 to 10 cm (3 to 4 in.)

DESCRIPTION: This is a fairly brightly colored mosaic darner. The male's face is pale green or yellow green, typically with a distinct black stripe and a relatively narrow T-spot stem with a straight-edged crossbar. The eyes are a dark blue green. The thorax is brown with relatively broad, yellow green lateral stripes, parallel sided but slightly wavy, and distinct yellow stripes on the front. The abdomen is black with fairly large, blue spots above, including atop segment 10. The blue areas do not get duller at colder temperatures. The male's cerci are paddle shaped with a spine on the lower tip extending beyond the top edge of the appendage. The male's pterostigma may be brown or black. The female has a similar pattern, or the pale areas on the abdomen are yellow and the eyes are brown. The female's cerci are spindle shaped, and the pterostigma is brown.

SIMILAR SPECIES: The male Shadow Darner *(A. umbrosa)* is most similar and often found with the male of this species. On perched individuals at close range, the unstriped face and reduced blue spotting on the abdomen above, including the lack of pale spots atop abdominal segment 10, might be seen on the Shadow Darner. Both sexes of Walker's Darner *(A. walkeri)* have a (typically) fainter, brown stripe between the frons and clypeus, whiter stripes on the thorax, very small or no pale markings on segment 10 above, and a black pterostigma. The male Variable Darner *(A. interrupta)* looks much darker on the wing. If the female Variable Darner has uninterrupted thoracic stripes, it can be difficult to distinguish, even in hand. The female Blue-eyed Darner *(A. multicolor)* lacks the black stripe on the face and has a tubercle under abdominal segment 1.

BEHAVIOR: Males on patrol stick to fairly short beats along shore or in channels in marshes. They tend to patrol a bit higher than other mosaic darners. Females oviposit in stems and leaves of emergent vegetation, up to about 1 m (3 to 4 ft) above the surface. Away from water, they forage along trails and in clearings, often coursing back and forth along a short route.

DISTRIBUTION: This species is fairly common in the Cascade-Sierran Province southward to Kern County, in the Northern Coastal Province southward to the San Francisco Bay Area, and in the Great Basin Province southward to Inyo County. An old record for San Diego County suggests it may occur in the mountains of southern California. It is found at sea level along the north coast and up to 2,600 m (8,500 ft) in the interior mountains. This species ranges widely in forested regions of western North America from southern Alaska to Arizona.

HABITAT: The Paddle-tailed Darner is found around marshy or boggy ponds, lakes, spring pools, and the backwaters of rivers and streams.

FLIGHT SEASON: It has been recorded flying mid-June through October.

WALKER'S DARNER   *Aeshna walkeri*

Pls. 18, 19, Fig. 8

LENGTH: 7 to 7.5 cm (3 in.); **WING SPAN:** 8.5 to 10 cm (3.5 to 4 in.)

DESCRIPTION: The male overall shows very little if any yellow or green color. The face is a pale, bone white with perhaps a light blue tint and a brown stripe in the suture between the frons and clypeus. The stem of the T-spot is thick, widening slightly at its base; the crossbar arms are angled back. The eyes are dark brown above, gray below. The back of the head is black. The thorax is dark brown with pale blue stripes; those on the sides are straight, bordered with black, and are white at their lower ends. The abdomen is black with fairly large, deep blue spots above. Segment 10 is mostly or entirely black above (it may have tiny, blue spots). The cerci are paddle shaped; the spine on the lower edge is short, not extending beyond the upper lobe of the appendage. The pterostigma is black. The female is similar to the male, but its abdominal spots are usually olive green (but may be blue). The female's cerci are tapered at the base.

SIMILAR SPECIES: All our other mosaic darners have the pterostigma brown (may be black on the male Paddle-tailed *[A. palmata]*). The Paddle-tailed Darner usually has a black stripe on its yellow green face, yellow stripes on the thorax, and more extensive pale spotting on the abdomen, including atop segment 10. The Shadow Darner *(A. umbrosa)* has green or yellow color in the lower part of the stripes on the side of the thorax, a brown pterostigma, and blue spots on the underside of the middle abdominal segments. An additional characteristic useful in hand is that the spines on the lower tip of the male cerci are longer on the male Paddle-tailed and Shadow Darners than on this species.

BEHAVIOR: Males patrol low over the water, meticulously checking the inlets and rocky crevasses lining the margins of pools in the streambed for females that have come to oviposit in seams of moss and the submerged roots of plants. Females and males not looking for mates forage along trails in and over the open brush on canyon slopes above the stream.

DISTRIBUTION: This species is almost entirely confined to California, ranging northward into western Oregon and southward into Baja California. It occurs nearly statewide in rugged country, typically at low to middle elevations, but with records to 1,800 m (6,000 ft). There are a few records for the western edge of the Great Basin and the Central Valley floor, but none from the Desert Province. It also breeds on Santa Cruz Island.

HABITAT: Walker's Darner is predictably found in season along rocky streams and rivers in arid foothill canyons. However, it has also been found on occasion at mountain meadows and lakes at higher elevations.

FLIGHT SEASON: Like many of our other darners, it is a late flyer, primarily from July through November, with a handful of May and June dates recorded.

Riffle Darners *(Oplonaeschna)*

The two species in this genus are found primarily in Mexico. One is found in the American Southwest in Arizona, Utah, and New Mexico, and it has been collected once in California.

RIFFLE DARNER *Oplonaeschna armata*
Pl. 17

LENGTH: 6.5 to 7.5 cm (2.5 to 3 in.); **WING SPAN:** 9 to 11 cm (3.5 to 4.5 in.)

DESCRIPTION: This is a large darner, mostly brown with blue, green, and yellow markings much like those of mosaic darners *(Aeshna)*. The face is yellow to light blue with a black cross stripe and a large, black, triangular T-spot on the frons. The male has dark bluish eyes, and the eyes of the female and the young male are olive to brown. The back of the head is tan. The anterior lateral stripe on the thorax is nearly pinched into two parts (a yellow green lower lobe and a blue upper segment) in its middle. The brown abdomen has small, blue spots above on the male; they may be green or yellow on the female. There is a thin, rearward-projecting lobe at the tip of segment 10 on the male. The male's cerci are long and paddlelike with a small but distinctive, toothed dorsal projection near the tip. The wings are clear with a short, dark pterostigma, and the radial sector does not branch in the outer wing.

SIMILAR SPECIES: The Riffle Darner looks like a mosaic darner at any distance. All those that might occur in southeastern California (Blue-eyed, California, Paddle-tailed, and Walker's Darners [*Aeshna multicolor, A. californica, A. palmata,* and *A. walkeri*]) have straight lateral stripes on the thorax and larger blue spots on the top of the abdomen. In-hand differences between this species and mosaic darners are noted above in the "Description" section.

BEHAVIOR: Males fly beats along intermittent streams.

DISTRIBUTION: An individual of this species, which ranges from Utah to Guatemala, was found in Water Canyon, Inyo County, in June. It may have been a wanderer, but the possibility remains that small populations exist in the remote and little-visited desert mountain ranges of southeastern California.

HABITAT: This species is found at intermittent streams in arid mountain canyons. These sites are mostly dry, consisting of disconnected pools and underground flows, except during the Southwest's late-summer rainy season, when flash floods pour through the streambeds.

FLIGHT SEASON: In Arizona, the Riffle Darner flies from June through August.

CLUBTAILS (Gomphidae)

This is an intriguing family of dragonflies, primitive in general appearance, with wide-set eyes, snakeskin-like patterns of light and dark markings, and expanded terminal segments on the abdomen, most striking on males, that produce the "clubbed tail" for which the family is named. They are in general wary and difficult to spot when perched because the intricate color patterns serve as excellent camouflage. The larvae are distinctive in appearance and designed for a life of burrowing in mud or sand (fig. 8). They typically inhabit streams and rivers; a few species also live in lakes. There are about 100 species in North America, most of them in the eastern United States, but some of the 12 species found in our state are restricted, or nearly so, to California and some of the neighboring states.

Grappletail *(Octogomphus)*

The sole member of this genus is a distinctive species restricted to the Pacific Coast of North America from British Columbia to Baja California.

GRAPPLETAIL   *Octogomphus specularis*
Pl. 20, Fig. 8
LENGTH: 5.25 cm (2 in.); **WING SPAN:** 6.5 cm (2.5 in.)
DESCRIPTION: This is a striking black and yellow or yellow green dragonfly. Pale colors fade from bright yellow to green to gray with age. Distinctive features include a mostly pale yellow or yellow green thorax with broad, black dorsolateral stripes bordering a middorsal yellow patch shaped like an inverted goblet; an abdomen that is mostly black above on segments 3 through 9 (with a hairline streak of yellow atop segments 3 through 7, which is scarcely visible in the field); and on males, a "club" consisting of an expanded segment 10 and large, multipronged abdominal appendages. An oval spot atop segment 10 and the top prongs of the cerci are bright yellow. A yellow dorsal stripe on the thorax between the wing bases extends onto the top of abdominal segments 1 and 2, with some yellow lateral markings on these segments as well. The face is yellow green striped with black, and the

eyes are gray. The legs and pterostigma are black. The female is similar to the male, but the abdomen is cylindrical without the clubbed tip and with a row of yellow dots and dashes (like Morse code) along the sides of the middle segments.

SIMILAR SPECIES: No odonate is very similar. The Pacific Clubtail *(Gomphus kurilis),* often found flying with this species, has two pale stripes on the front of the thorax, as well as a black stripe on the side.

BEHAVIOR: Mature males may be found on sunny rocks along stream banks, females in willows and other brush and low trees some distance from water. Females and nonbreeding males forage away from water at spring seeps on the ground or surface vegetation or from perches in open brush. Females visit shade-dappled pools and oviposit by rapidly tapping the abdomen tip in the water a few times at various spots, then quickly depart.

DISTRIBUTION: This species is fairly common and widely distributed over the length of the state, primarily west of the crest of the Sierra Nevada and the southern deserts, below 1,500 m (5,000 ft). There are a few records for the Central Valley, none south of Stockton. It is occasionally found in the mountains east of the Pacific Crest (Eagle Lake, Lassen County, and the headwaters of the Mojave River, San Bernardino County).

HABITAT: This species inhabits creeks and small rivers, especially where these have appreciable slope and a mix of rock-strewn riffles and pools. When found away from water, it is usually in brushy clearings, at seep springs, and along trails in wooded country.

FLIGHT SEASON: The Grappletail is primarily seen flying from April through August (there is one October record). Recent emergers are most common at lower elevations in spring and fly upstream to breed in June and July at higher elevations.

Common Clubtails *(Gomphus)*

This is a large, diverse group, split by some into a number of genera or subgenera. Thirty-eight of the approximately 50 species in the genus are found in North America, and most of these occur east of the Rockies. Only one species is found in California.

PACIFIC CLUBTAIL   *Gomphus kurilis*
Pl. 20, Fig. 8
LENGTH: 5.25 cm (2 in.); **WING SPAN:** 6 to 7 cm (2.5 to 3 in.)

DESCRIPTION: This clubtail looks mostly black with pale green and yellow markings. The face is pale yellow green, separated from the yellow occiput by a black vertex. The eyes are blue gray. The thorax is dull, light green broadly striped with black or dark brown, as follows: a central stripe on the front; two dorsolateral stripes (with a pale, hairline slash down the middle of each); and a single black, diagonal stripe on each side. The legs are black. The abdomen is mostly black with a dorsal and two lateral yellow stripes on segments 1 and 2 that reduce to spots of variable size along the sides and long, narrow triangles atop segments 3 through 7. Segments 8 and 9 are slightly expanded into a club, with large, bright yellow lateral patches. The dorsal yellow spots atop segments 8 and 9 are variable in size and sometimes absent on segment 9. Segment 10 is black above. The female is similar, but the abdomen is less club tipped and has yellow patches that are somewhat more extensive. The thorax is brighter on younger individuals, and the pale areas are yellow.

SIMILAR SPECIES: The Grappletail *(Octogomphus specularis)* has a single, central, yellow stripe on the front; lacks the black diagonal stripe on the side of the thorax; and has a mostly black abdomen above. The Olive Clubtail *(Stylurus olivaceus)* and snaketails *(Ophiogomphus)* have the sides of the thorax olive gray and bright green, respectively, without the dark diagonal stripe, and have a pale stripe on the femur of the hind leg.

BEHAVIOR: At breeding sites in the late morning and afternoon, males perch on low vegetation, rocks, or logs near shore or the beach itself. Away from water, males and females can be found along trails and other bare patches of ground or perched low in brush or weeds. They often flush with a distinctive bouncing, roller coaster–like flight. The female oviposits alone, hovering over water and tapping the surface with her abdomen a few times at various spots.

DISTRIBUTION: Virtually the entire range of this species is restricted to two states, California and Oregon, with a few populations in Washington and Nevada. Within California, it is widely distributed and fairly common in the northern half of the state southward to Monterey and Madera Counties. It is found from sea level to 1,800 m (6,000 ft) at mountain lakes.

HABITAT: Found at a greater variety of aquatic sites than our other clubtails, it breeds in rivers, streams, sloughs, ponds, and lakes, primarily with mud or silt bottoms. It is found away from water in open brush, riparian woods, and orchards.

FLIGHT SEASON: This species has an early season, flying from April through July (occasionally into August).

Hanging Clubtails *(Stylurus)*

Most species in this genus are similar in general form to those in the genus *Gomphus,* and they were at one time included in that group. They differ from our other clubtails, however, in that they often perch by hanging from vegetation, even high in trees, and less frequently land on the ground. Males patrol over the open water of rivers and streams in search of females. Captured females are taken up into nearby brush or trees for mating. Females deposit eggs, unattended, by dipping the abdomen into the water while in flight. Of the 11 North American species in the genus, three occur in California.

OLIVE CLUBTAIL   *Stylurus olivaceus*
Pl. 21

LENGTH: 5.5 to 6 cm (2 to 2.5 in.); **WING SPAN:** 8 cm (3 in.)

DESCRIPTION: This is an elegantly patterned but relatively colorless dragonfly. Most distinctive is the thorax pattern: The front is dark brown or black with a pale olive gray, U-shaped mark, and the sides are unmarked, pale olive gray. The face is a dull gray brown with a black band across the vertex. The eyes are blue gray. The abdomen is mostly black with dorsal gray patches atop segments 3 through 7 resembling small "darts," aimed rearward, and has pale areas low on the sides of these segments, often with small black spots at their posterior ends. On the male, segments 7 through 9 are laterally expanded to form a flattened club with conspicuous, pale yellow lateral patches and a large, pale oval atop segment 9; a smaller pale spot is atop segment 8. The femur is mostly pale gray with a black stripe, and the tibia and tarsus are black. The pterostigma is brown. The female is patterned like the male but without a pronounced club. Some Great Basin populations have pale areas that are slightly yellower and more extensive.

SIMILAR SPECIES: The Pacific Clubtail *(Gomphus kurilis)* has a broad, black diagonal stripe on the side of the yellow green thorax, and all black legs. Snaketails *(Ophiogomphus)* have the pale areas of the thorax a brighter green, and large, yellow patches atop the middle abdominal segments. The Brimstone Clubtail *(S. intricatus)* is smaller and more extensively yellow, with limited black markings.

BEHAVIOR: Breeding males patrol low in zigzag flight over deep, muddy water away from shore. They hang at roosts in overhanging willows and other trees and brush along the water's edge. Females and immature males feed away from water in open riparian woods. They are difficult to find when perched because of their dull, cryptic colors and their habit of perching high in trees and large shrubs, often in the shade.

DISTRIBUTION: Of seemingly local occurrence in northern and central California, this species is easily overlooked and is probably more widespread than the records suggest. It has been found along some of the rivers and creeks of the Central Valley and surrounding foothills from Glenn County southward to Stanislaus and Mariposa Counties and along the rivers of the Great Basin Province from Modoc County southward to Inyo County. There is one record for the central coast at Santa Cruz, Santa Cruz County. Records from west of the Sierras are all below 600 m (2,000 ft), but on the eastern slope it occurs at about 900 to 1,400 m (3,000 to 4,500 ft).

HABITAT: The Olive Clubtail is typically found along mud-bottomed rivers and larger streams, foraging over adjacent riparian woods and clearings. It occasionally wanders a bit further afield.

FLIGHT SEASON: This species has a brief, late-summer season, from late June to early September, with most records in July.

RUSSET-TIPPED CLUBTAIL — *Stylurus plagiatus*
Pl. 21

LENGTH: 5.5 to 6.5 cm (2 to 2.5 in.); **WING SPAN:** 7 to 8 cm (3 in.)

DESCRIPTION: This is a large, handsome clubtail. Its eyes are turquoise blue. The thorax is mostly green with dark brown stripes on the front and sides. The abdomen is black with creamy yellow markings, including thin, pale "darts" atop segments 3 through 7. Segments 7 through 9 of the male widen to form a "club"; segments 8 and 9 are mostly orange yellow with a dark brown dorsal patch on the rear margin. Segment 10 and the terminal appendages are dark brown to black above. The pterostigma is dark brown. The femur is black with a pale stripe; the tibia and tarsus are black. The female has a similar pattern but has a slender, cylindrical abdomen that lacks a clubbed tip.

SIMILAR SPECIES: There are few other clubtails in its range in California. The Brimstone Clubtail *(S. intricatus)* has a mostly yellow green thorax with minimal brown striping, more yellow on the abdomen, and a pale stripe on the tibia. The White-belted Ringtail *(Erpetogomphus compositus)* has white bands on the abdomen and has pale cerci. The Gray Sanddragon *(Progomphus borealis)* has a gray and brown striped thorax and long, pale cerci.

BEHAVIOR: Males patrol low over water in the early morning and retire to roost and mate with females atop bushes nearby in late morning. Flight away from water is swift and direct. They hang to roost in shady spots in desert scrub within 3 m (10 ft) above ground level, where they are hard to see before they fly.

DISTRIBUTION: Widely distributed in the eastern United States and the Southwest, the Russet-tipped Clubtail barely enters California in Imperial and Riverside Counties, occurring along the Colorado River and in the Imperial Valley at elevations barely above sea level.

HABITAT: This species is found along deep rivers and irrigation canals with sand or silt bottoms. Away from water, it forages over desert scrub and agricultural lands.

FLIGHT SEASON: Dates range from July to September.

BRIMSTONE CLUBTAIL *Stylurus intricatus*
Pl. 21

LENGTH: 4 to 5.5 cm (1.5 to 2 in.); **WING SPAN:** 5.5 to 6.5 cm (2 to 2.5 in.)

DESCRIPTION: This desert dweller is smaller and paler yellow overall than our other hanging clubtails. The face is yellow, and the eyes are pale gray. The thorax is pale greenish yellow with variable thin brown striping on the front and upper sides. The abdomen is mostly yellow, paler and green tinged on segments 1 through 6, brighter on segments 7 through 10; segments 7 through 9 expand into a modest club on males. There are dark abdominal marks limited to narrow, black apical rings on segments 3 through 6, and a row of black or brown spots, two per segment, on segments 3 through 7. Segments 8 through 10 are mostly sulfur yellow; segment 8 and sometimes segment 9 have a dark brown dorsal spot. The terminal appendages are yellow with black tips. The pale areas on the legs include most of the femur and a yellow stripe on the tibia. The pterostigma is yellow to tan. The female is patterned much like the male, but the abdomen is cylindrical, not clubbed at the tip.

SIMILAR SPECIES: The Pale Snaketail *(Ophiogomphus severus)* has dark stripes on the sides of the abdomen, including segments 8 and 9. It typically perches on the ground, and its known range in California does not overlap with this species. Other hanging clubtails are larger and have black tarsi and extensive dark markings on the thorax and the abdomen. The Western Pondhawk *(Erythemis collocata)* female has eyes in contact with each other and a black dorsal stripe on the abdomen.

BEHAVIOR: Brimstone Clubtails perch in low brush and weeds when foraging away from water, on driftwood and streamside brush at breeding sites, and occasionally on the ground.

DISTRIBUTION: This pale clubtail inhabits arid, open country in the western United States. In California, it has been found in the Owens River valley of Inyo County at about 1,200 m (4,000 ft) and at scattered locations near sea level in the Desert Province of Riverside and Imperial Counties.

HABITAT: Breeding occurs in muddy, alkaline rivers and irrigation canals in open desert country.

FLIGHT SEASON: This is a late flyer in California, recorded dates ranging from July through September.

Snaketails *(Ophiogomphus)*

The colorful snaketails have a green thorax with variable brown stripes and the abdomen is patterned like a blotched snakeskin in yellow, black, and white. The females have the "tail club" much less pronounced than that of males. They typically occur along streams and rivers, occasionally at lakes, where males in breeding condition sit out on sunny rocks and gravel bars from which they make low flights out over the water. Females oviposit by flying low over riffles and tapping the tip of the abdomen at various spots on the water's surface. Away from water they perch on the ground, weeds, or low brush in open areas. Eighteen species are found in North America, the rest in Eurasia. Four rather similar species occur in California. They cannot be told apart except at close range or in the hand.

PALE SNAKETAIL   *Ophiogomphus severus*
Pl. 22

LENGTH: 5 cm (2 in.); **WING SPAN:** 6 to 6.5 cm (2.5 in.)

DESCRIPTION: This is our palest snaketail, especially on the thorax. The face is yellow, the vertex is black or mostly so with perhaps a small yellow patch near the yellow occiput, and the eyes are gray. The thorax is yellow green aging to yellow olive. It has a brown oval spot on either side of a thin, brown stripe on the midline of the front of the thorax, and another thin, brown stripe on either side just below each oval spot. The abdomen is mostly yellow above and white on the lower sides with a jagged black lateral stripe (occasionally broken into dark spots, one per segment) on segments 2 through 9. Segments 7 through 9 are expanded into a club, and segment 10 is mostly yellow with a faint brown stripe on each side. The tibia has a pale stripe. The pterostigma is gray. The male has stout cerci about the same length or slightly longer than the epiproct. The female lacks occipital horns.

SIMILAR SPECIES: The Great Basin Snaketail (*O. morrisoni*) can look very similar, but it usually has dark brown stripes on the front of the thorax, one down the center and two on either side. Any snaketail outside of the known range of the Pale Snaketail in California with isolated, dark, short stripes or ovals on the front of the thorax needs to be examined in hand. Check the shape and relative lengths of the cerci and epiproct (females may be impossible to separate). The Sinuous Snaketail (*O. occidentis*) has wavy-margined dorsolateral stripes on the thorax, an upturned epiproct on the male, and postoccipital horns on the female. The Brimstone Clubtail (*Stylurus intricatus*) lacks black stripes on the side of the abdominal segments, and its range does not overlap this species in California.

BEHAVIOR: Males perch on the ground and rocks at streamside in the morning. Often the abdomen is raised in the obelisk position as the weather warms up. Foraging individuals perch atop low brush nearby.

DISTRIBUTION: Although widespread in the interior of western North America, this snaketail has been found in California only in eastern Modoc County (Surprise Valley, Alturas) at elevations of 1,300 to 1,400 m (4,300 to 4,600 ft).

HABITAT: The Pale Snaketail is found at small, sand-bottomed creeks and spring runs in sagebrush country.

FLIGHT SEASON: All California records are from June and July.

GREAT BASIN SNAKETAIL *Ophiogomphus morrisoni*
Pl. 22
LENGTH: 5 to 5.25 cm (2 in.); **WING SPAN:** 6.5 cm (2.5 in.)

DESCRIPTION: This is a somewhat variable dragonfly, generally larger and paler in desert lowlands (subspecies *O. m. nevadensis*), and smaller and darker at mountain lakes. The face is yellow green, and the vertex black with a central, yellow, oval spot. The eyes often are bright aqua blue or blue gray. The thorax varies from gray green to pale lime green, with five relatively narrow, black to brown stripes on the front, one in the middle and a pair on either side, the latter not wavy and separated by a green stripe of variable width (from a hairline to about the same width as the brown stripes). Occasionally the lateral stripes are reduced to a stripe and an oval spot, as on the Pale Snaketail *(O. severus)*. The abdominal pattern is similar to that of other clubtails, at the dark extreme more like the Bison Snaketail *(O. bison)*, at the light extreme more like the Sinuous Snaketail *(O. occidentis)*. The tibia has a pale stripe. The pterostigma is brown or gray brown. The male's cerci are short and thick, about the same length or shorter than the epiproct. The female is often without occipital horns but may have a widely spaced pair on the top edge of the occiput.

SIMILAR SPECIES: See the description of the Pale Snaketail, which typically has the front of the thorax green with brown, oval spots. The Sinuous Snaketail has wavy brown or black dorsolateral stripes on the thorax, cerci longer than the upturned epiproct on the male, and both occipital and postoccipital horns on the female. The Bison Snaketail has a single, broad, dark, dorsolateral stripe on each side of the thorax with, at most, an incomplete green hairline; the cerci are thin and longer than the epiproct, and the female has long, narrow occipital horns.

BEHAVIOR: They behave similarly to other snaketails. Males perch on rocks or gravel beds along the shore, making short patrols out over the stream or lake. Females oviposit like other snaketails in riffles or the wave-washed shorelines of lakes.

DISTRIBUTION: Although this species does occur along streams and rivers in the Great Basin Province from Lassen County southward to Inyo County as its name suggests, it is also found along rivers and at high mountain lakes in the Cascade-Sierran Province from Siskiyou County southward to Tulare County. Recorded elevations range from 600 m (2,000 ft) in the north to 3,000 m (10,000 ft) in the southern Sierra Nevada. A specimen from the Central Valley (Sacramento) probably represents a stray from higher elevations.

HABITAT: The gravelly margins of creeks, rivers, and lakes are frequented by this snaketail.

FLIGHT SEASON: This is a midsummer species, with flight dates from late May into August.

SINUOUS SNAKETAIL   *Ophiogomphus occidentis*
Pl. 22

LENGTH: 5 cm (2 in.); **WING SPAN:** 6 to 6.5 cm (2.5 in.)

DESCRIPTION: The face is yellow to yellow green, the vertex black with a yellow patch behind the simple eyes. The compound eyes are gray, darker above. Populations in and around the Central Valley (the subspecies *O. o. californicus*) have more extensive pale areas and paler brown stripes on the thorax. The thorax is light green to yellow green and has five dark stripes on the front like the Great Basin Snaketail *(O. morrisoni),* but the uppermost lateral stripe is wavier, slightly S-shaped, and separated from the lower lateral stripe by a thin, wavy, green line (may be a broken hairline or, on *O. o. californicus,* nearly as wide as the dark stripes it separates). The pattern on the abdomen is similar to that of other snaketails, at the dark extreme like the Bison Snaketail, at the light extreme like the Great Basin Snaketail *(O. morrisoni nevadensis).* Abdominal segment 10 and the terminal appendages are largely yellow. There is a narrow, pale stripe on the back of the tibia, and the pterostigma is brown. The male cerci are fairly narrow, extending beyond the epiproct, which curls rather sharply upward at the tip. The female is patterned like the male and has a pair of horns on top of the occiput as well as two small, widely spaced postoccipital horns, the latter sharply hooked outward.

SIMILAR SPECIES: The Bison and Great Basin snaketails have straighter, less sinuous brown stripes on the upper side of the thorax. In hand, none of our other snaketails have the strongly upturned epiproct on the male or both occipital and postoccipital horns on the female.

BEHAVIOR: Breeding males sit on the stream bank like other snaketails. Nonbreeding individuals move away from water to forage in open, grassy areas, where they perch on the ground or low weeds and sally out to catch prey.

DISTRIBUTION: A species of local occurrence, the Sinuous Snaketail is often quite numerous where found emerging. It ranges through the Pacific Northwest from British Columbia to northern California, where it occurs in the California and Cascade-Sierran Provinces (including the Pit River drainage eastward to the edge of the Great Basin) southward to Sacramento County. Primarily a valley and foothill species, it has been recorded from near sea level up to 3,100 m (4,300 ft).

HABITAT: The English name of this species refers to the wavy, dark stripes on the thorax but also suggests the preferred habitat: major rivers and larger streams in their lower, more meandering reaches. Sandy, muddy, or gravel-bottomed sites are occupied. It is occasionally found along lakeshores and may move some distance from water to forage in grasslands or open woods.

FLIGHT SEASON: The Sinuous Snaketail flies rather early, from late March into August.

BISON SNAKETAIL   *Ophiogomphus bison*
Pl. 22

LENGTH: 5 cm (2 in.); **WING SPAN:** 6 to 6.5 cm (2.5 in.)

DESCRIPTION: This is our darkest snaketail. The face is yellow green, the vertex black, the eyes gray. The thorax is grass green with three broad, dark brown stripes on the front, one in the middle and two on the upper sides. There is often a faint hairline of green in the dark lateral stripes. The abdomen is mostly black, with yellow, triangular to oval patches atop segments 2 through 10 and white areas low on the sides of segments 1 through 6; segments 7 through 9 are expanded into a club with yellow lateral patches. The tibia is all black, as is the pterostigma. The male's cerci are relatively narrow, longer than the epiproct. The epiproct is nearly straight in profile. The female has two fairly long, narrow occipital horns projecting from midocciput (the "bison horns").

SIMILAR SPECIES: Other snaketails have each dorsolateral dark stripe on the thorax split into two stripes or reduced to an oval patch and a thin stripe. In hand, note differences in the male abdominal appendages and the occipital structure of the female. The Pacific Clubtail *(Gomphus kurilis)* has a fairly broad black stripe on the side of the thorax.

BEHAVIOR: Males perch on gravel bars, where they are hard to see until flushed. They patrol out over riffles and pools with low, zigzag flight. Females and nonbreeding males forage away from the water in open areas, usually from a low perch in the weeds or brush.

DISTRIBUTION: This snaketail is found in the foothills and lower mountain zones of the Coast Ranges and Cascade-Sierran Province from the Oregon border southward to Monterey and Tulare Counties. It occurs near sea level on the coast and up to 1,500 m (5,000 ft) in the Sierra Nevada. It is of rare occurrence in the Sacramento Valley southward to Sacramento. Outside of California, it is known only from Oregon, Nevada, and Utah.

HABITAT: Breeding occurs along gravel bars with scattered willows and other brush and debris on fast-flowing creeks and small rivers. It is sometimes found along irrigation ditches and foraging within 100 m (300 ft) or so from water.

FLIGHT SEASON: This species is found on the wing from April through October but is most common May through July.

Ringtails *(Erpetogomphus)*

The alert and attractive ringtails comprise about 20 stream-dwelling species restricted to the Americas, six found north of Mexico. Two southwestern species occur in California. The English name refers to their most distinctive feature: bright white rings on the bases of the middle abdominal segments.

WHITE-BELTED RINGTAIL *Erpetogomphus compositus*
Pl. 23

LENGTH: 4.5 to 5.5 cm (2 in.); **WING SPAN:** 7 cm (3 in.)

DESCRIPTION: This is a striking dragonfly that looks like a colorful composite of different species. The face is a creamy white, the eyes a pale, blue gray. The thorax is striped with brown between alternating bands of yellow green and white. The abdomen has vivid white rings at the bases of the black, middle abdominal segments, and some small, pale marks atop these segments. The male's abdomen is distinctly clubbed; segments 8 through 10 look mostly bright, tawny yellow in the field, with some poorly defined, dark brown patches on their upper surfaces. The cerci are yellow and taper to a smooth, slightly downcast point. The femur is pale with a black stripe, and the tibia is black. The female has a similar pattern on the thorax, but the abdomen is not clubbed and the occiput has a narrow, wavy dorsal ridge.

SIMILAR SPECIES: The Serpent Ringtail *(E. lampropeltis)* is duller overall, the thorax striped brown and gray. Snaketails lack the bright white rings on the middle abdominal segments and have the lower sides of the thorax green, with at most one thin, brown stripe.

BEHAVIOR: Males perch in exposed, sunny spots on rocks, gravel bars, or low vegetation in midstream or on the shore, typically near riffles. They patrol with a low, hopping flight over water, chasing other males, even harassing other species (e.g., patrolling clubskimmers *[Brechmorhoga]*). Females and nonbreeding males foraging away from water can be found perched on tops of bushes, fence wires, even up in trees.

DISTRIBUTION: This species ranges southward from Shasta County in the arid valleys and foothills of the California Province, in the Desert Province, and from Mono County southward in the Great Basin Province. They are typically found at lower elevations but have been recorded at above 1,500 m (5,000 ft) in the Great Basin east of the Sierra Nevada.

HABITAT: It occurs along rivers, streams, and irrigation ditches; fast-moving water with riffles and pools, beds of sand and gravel, and rocky shoreline are preferred. This species forages away from water in arid brushlands and grasslands with scattered trees.

FLIGHT SEASON: The White-belted Ringtail is on the wing from April to October.

SERPENT RINGTAIL   *Erpetogomphus lampropeltis*
Pl. 23, Fig. 8
LENGTH: 4 to 5.5 cm (1.5 to 2 in.); **WING SPAN:** 7 cm (3 in.)
DESCRIPTION: The California subspecies *(E. l. lampropeltis)* is like a duller version of the White-belted Ringtail *(E. compositus)*. The thorax is striped brown and gray or gray green. The abdomen is black with white rings at the bases of the middle abdominal segments; some narrow, yellow markings are on the top and lower sides of these segments. Segments 8 and 9 of the male are expanded into a club that is dark brown on top and orange yellow on the sides, with thin, pale yellow rings on the posterior edge of segments 7 through 10. The male's cerci are yellow and "pinched" at the tip. The female's occiput has a dark ridge along its hind margin.

SIMILAR SPECIES: The White-belted Ringtail has a green-, white-, and brown-striped thorax; the male's abdominal club is paler, and orange yellow in color. The Gray Sanddragon *(Progomphus borealis)* lacks white rings on the abdomen.

BEHAVIOR: Males perch on rocks in midstream and near shore. They are somewhat flighty but often allow close approach. They may perch on exposed twigs on brushy hillsides away from water.

DISTRIBUTION: This subspecies is known only from the mountains of the coastal slope of southern California, from Ventura, Los Angeles, and southwestern San Bernardino Counties southward to San Diego County. It has been found at elevations ranging from around 180 to 800 m (600 to 2,500 ft). Other populations occur from Arizona and Texas southward to Guatemala.

HABITAT: This Serpent Ringtail subspecies occurs along permanent, usually spring-fed streams and rivers in foothill country. Away from water it may be found in open brush and weedy areas.

FLIGHT SEASON: It flies late, from mid-June into October.

Sanddragons *(Progomphus)*

This large genus of over 65 primarily Neotropical, stream-dwelling dragonfly species is represented in North America by just four species, only one of which occurs in California.

GRAY SANDDRAGON   *Progomphus borealis*
Pl. 23

LENGTH: 5.5 cm (2 in.); **WING SPAN:** 7 cm (3 in.)

DESCRIPTION: This is a fairly dark dragonfly, subtly patterned. The face is gray green, the top of the frons is tawny yellow. The vertex is black with a large, yellow, oval patch behind the simple eyes. The occiput is yellow, the eyes gray to dark olive gray. The thorax is rusty brown in front with an elaborate yellow pattern in the shape of a harp; the sides are gray with a brown stripe below the base of the hind wing. There is a yellow, horizontal band across the top of the thorax between the dark fore and hind wing bases. The femur is black with a pale stripe, and the tibia is black. The slender abdomen is mostly black with narrow, creamy yellow triangles, pointed rearward, atop segments 3 through 7; a small, yellow, basal spot tops segment 8; and small, yellow or rusty spots are on the sides of segments 9 and 10. Segments 8 and 9 are slightly expanded to form a modest club. On the male, the cerci are long and bright creamy white or yellow, and the epiproct, which is black and divided in two, is unique among our dragonflies. The pterostigma is black. The female resembles the male in pattern but has only a faint hint of a club and abdominal segment 10 is mostly yellow.

SIMILAR SPECIES: The Olive Clubtail *(Stylurus olivaceus)* is plain olive gray on the side of the thorax and hangs in trees or brush. Ringtails *(Erpetogomphus)* have white bands on the abdominal segments and a fairly large, pale "club" on the male. Our other clubtails have extensive yellow or green areas on the thorax.

BEHAVIOR: Males sit in open, sunny spots on the shoreline or on driftwood in the water, making occasional sallies for food or to chase other males. They are in general quite wary but at times allow close approach.

DISTRIBUTION: This is a species of the more arid foothills and valleys of the California Province southward from Sonoma, Lake, and Shasta Counties, the Desert Province, and the Great Basin Province from eastern Lassen County southward, occasionally ranging into the mountains of these regions to about 1,300 m (4,300 ft). Not at all a boreal species as the scientific name implies, it does, however, have a more northerly distribution than most other sanddragon species, ranging from southern Idaho to Mexico.

HABITAT: The Gray Sanddragon is found along sand-bottomed rivers and creeks where the larva burrows long tunnels just below the surface of the sandy beds. It does not stray far from water but occasionally occupies exposed perches on low weeds in open areas.

FLIGHT SEASON: Recorded dates cover a rather broad span, from late March (in the south of the state) to late August.

SPIKETAILS (Cordulegastridae)

These large, striking dragonflies are all the more impressive because they seem to go about their lives fearlessly, oblivious to much of what is going on around them, including human activity. Perhaps because they are so conspicuous, they have been given a number of evocative common names such as "biddie" and "flying adder." The English name used here refers to the spikelike ovipositor of females. A nearly cosmopolitan group, the Cordulegastridae is represented in North America by eight species in the genus *Cordulegaster,* one of which occurs in California.

PACIFIC SPIKETAIL   *Cordulegaster dorsalis*
Pl. 24, Figs. 8, 11
LENGTH: 7 to 8.5 cm (3 to 3.5 in.); **WING SPAN:** 9 to 10.5 cm (3.5 to 4 in.)

DESCRIPTION: This is a huge, black dragonfly with yellow thoracic stripes (a pair on top and a pair on either side) and yellow saddles on the abdominal segments. The black legs are relatively short. Adults have brilliant, aqua blue eyes that barely touch at the top of the head. The ovipositor of the female is long and trowel shaped. A Great Basin form *(C. d. deserticola)* has more extensive pale markings. Big, ragged teeth on the palps of the cup-shaped labial mask give the bulky larva a distinctive facial expression.

SIMILAR SPECIES: Two other large dragonflies with similar patterns are the Black Petaltail *(Tanypteryx hageni)* and the Western River Cruiser *(Macromia magnifica).* Both species occur with the spiketail at some sites in the North Coast Ranges and Cascade-Sierran Province. The Black Petaltail doesn't patrol long, low beats along streams and roads (although it perches on the latter and other flat surfaces) and has a distinctly different pattern of yellow spots on the thorax and abdomen. At close range, note the petaltail's extremely long and slender pterostigma and blackish compound eyes that don't touch each other. The Western River Cruiser has gray green eyes in fairly broad contact and a single, pale yellow stripe on each side of the brown thorax. It typically patrols larger streams and rivers.

BEHAVIOR: Males patrol long, low routes along watercourses, searching for females. Almost always seen in flight, they occasionally hang low in streamside vegetation. Females seldom visit water except to breed and oviposit. The latter is accomplished alone by hovering vertically over shallow water and repeatedly plunging the ovipositor into the muddy or sandy substrate. Away from water, Pacific Spiketails forage along linear routes and are frequently seen flying great distances down mountain roads a few feet off the ground, seemingly indifferent to hikers or passing vehicles. Their flight path is rather straight and steady, with occasional acrobatic deviations to catch prey.

DISTRIBUTION: This species is found nearly statewide in wooded country but is lacking from much of the Central Valley and Desert Province. The desert subspecies occurs at springs in more open country in Inyo County. Recorded elevations range from sea level to a little over 1,800 m (6,000 ft).

HABITAT: The Pacific Spiketail breeds along headwater streams, small, permanent boggy rivulets, and spring runs. The large larva lies concealed by surface muck and ambushes prey.

FLIGHT SEASON: This species flies mostly in summer, from mid-May to early October.

PLATES

PLATE 1   Broad-winged Damsels

Large damselflies easily distinguished by the patches of color on the wings.

American Rubyspot
(Hetaerina americana)   PAGE 40

Length: 4 to 4.5 cm (1.5 in.)
Wing span: 6.5 cm (2.5 in.)

The male has striking red patches at the base of the wings. The female's wing bases are dull amber.

River Jewelwing
(Calopteryx aequabilis)   PAGE 39

Length: 4.5 to 5 cm (2 in.)
Wing span: 6.5 to 7 cm (2.5 to 3 in.)

The male is metallic green and has distinctive black wing tips. The wings of the female are smoky with dark brown tips and contrasting white pterostigmata.

PLATE 1   Broad-winged Damsels

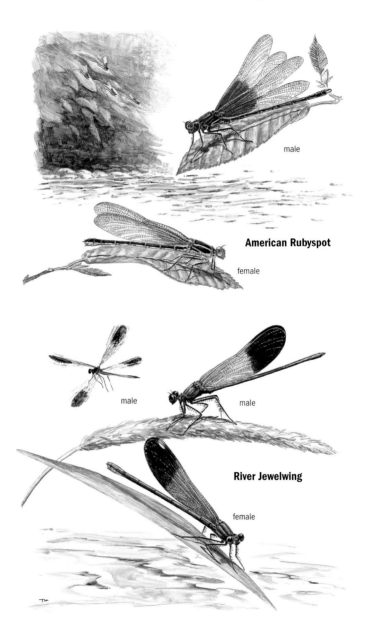

male

American Rubyspot

female

male

male

River Jewelwing

female

PLATE 2   **Stream Spreadwings**

Large damselflies that hang from vegetation with their wings spread.

California Spreadwing
(Archilestes californica)   PAGE 41

Length: 4.5 to 6 cm (2 to 2.5 in.)
Wing span: 5.5 to 7 cm (2 to 3 in.)

The thorax is brown and black above, with an incomplete white side stripe. The mature male typically has pale, golden tan pterostigmata. The male's paraprocts are more or less parallel.

Great Spreadwing
(Archilestes grandis)   PAGE 43

Length: 5 to 6.25 cm (2 to 2.5 in.)
Wing span: 6.5 to 8 cm (2.5 to 3 in.)

This is the largest damselfly in California. The thorax and abdomen are dark metallic green above in mature individuals, with a complete yellow side stripe on the thorax. The mature male has a dark pterostigma, and the paraprocts diverge (i.e., they are splayed out to the sides).

PLATE 2   Stream Spreadwings

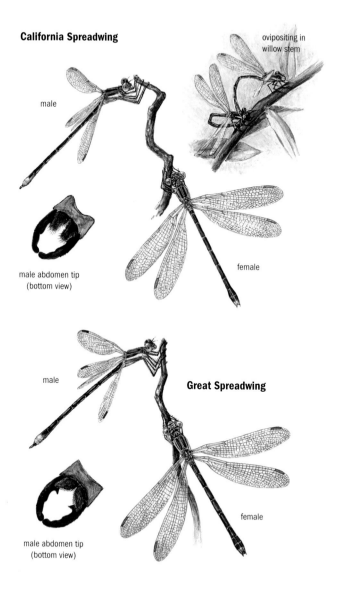

California Spreadwing

ovipositing in willow stem

male

male abdomen tip
(bottom view)

female

male

Great Spreadwing

male abdomen tip
(bottom view)

female

PLATE 3   Pond Spreadwings, Males

Midsized damselflies with blue eyes, abdominal segments 9 and 10 pale gray above, and wings spread when perched.

Spotted Spreadwing
(Lestes congener)   PAGE 44

Length: 3.5 to 4 cm (1.5 in.)
Wing span: 4.5 cm (2 in.)

This damselfly is black above and white below, with four black spots on the underside of the thorax. The paraprocts are short and stubby.

Emerald Spreadwing *(Lestes dryas)* PAGE 46

Length: 3 to 4 cm (1 to 1.5 in.)
Wing span: 3.5 to 4.5 cm (1.5 to 2 in.)

A stocky spreadwing, metallic green above and with paraprocts like those of the Black Spreadwing.

Black Spreadwing *(Lestes stultus)*   PAGE 45

Length: 3.5 to 4.5 cm (1.5 to 2 in.) Wing span: 5 cm (2 in.)

This spreadwing is similar to the Spotted Spreadwing but is typically larger and stockier. The top of the thorax has variable purple, bronze, or dull green iridescence. The abdomen is metallic green above. The paraprocts are of moderate length and distinctly foot shaped, the "toes" pointing in.

Common Spreadwing
(Lestes disjunctus)   PAGE 47

Length: 3.5 to 4 cm (1.5 in.) Wing span: 4.5 cm (2 in.)

The thorax is prunose blue-gray. The paraprocts are long and thin; they are typically held parallel but may be crossed (compare to the Lyre-tipped Spreadwing).

Lyre-tipped Spreadwing
(Lestes unguiculatus)   PAGE 48

Length: 3 to 4 cm (1 to 1.5 in)
Wing span: 3.5 to 4.5 cm (1.5 to 2 in.)

The thorax is bronze and green above. The curvy paraprocts form a distinctive "lyre" shape. Note white borders on the pterostigma.

PLATE 3   Pond Spreadwings, Males

Male Pond Spreadwings

abdomen tip
(top view)

**Emerald
Spreadwing**

**Spotted
Spreadwing**

abdomen tip
(top view)

Common Spreadwing

**Black
Spreadwing**

abdomen tip
(top view)

abdomen tip
(top view)

abdomen tips
(top views)

Lyre-tipped Spreadwing

PLATE 4   Pond Spreadwings, Females

See range maps on plate 3

Emerald Spreadwing *(Lestes dryas)*   PAGE 46

Length: 3 to 4 cm (1 to 1.5 in.) Wing span: 3.5 to 4.5 cm (1.5 to 2 in.)

The mature female usually has a metallic green thorax above. If her thorax is brown or bronze above, she is more or less identical to the Black
Spreadwing. The abdomen is relatively short and stubby, the ovipositor
as long as or longer than the length of abdominal segment 7.

Black Spreadwing *(Lestes stultus)*   PAGE 45

Length: 3.5 to 4.5 cm (1.5 to 2 in.) Wing span: 5 cm (2 in.)

This female is large and stocky. The thorax is black above, with purple,
bronze, or green iridescence. The ovipositor is as long as or longer than
the length of abdominal segment 7.

Spotted Spreadwing *(Lestes congener)*   PAGE 44

Length: 3.5 to 4 cm (1.5 in.) Wing span: 4.5 cm (2 in.)

The thorax is black above and creamy white with four black spots on the
underside. Abdominal segment 7 is longer than the ovipositor. The teneral is salmon pink and bronze (tenerals of other spreadwings are similar).

Common Spreadwing *(Lestes disjunctus)*   PAGE 47

Length: 3.5 to 4 cm (1.5 in.) Wing span: 4.5 cm (2 in.)

This female is similar to the female Spotted Spreadwing, but has thicker
pale stripes on the upper sides of the thorax and lacks the four dark spots
on the underside. The back of the head and the pterostigmata are dark.

Lyre-tipped Spreadwing *(Lestes unguiculatus)*   PAGE 48

Length: 3 to 4 cm (1 to 1.5 in.) Wing span: 3.5 to 4.5 cm (1.5 to 2 in.)

The thorax is dark brown above and creamy yellow below, with broad
white stripes on the upper sides. The rear of the head is pale and there are
narrow white stripes on either end of the pterostigma.

PLATE 4   Pond Spreadwings, Females

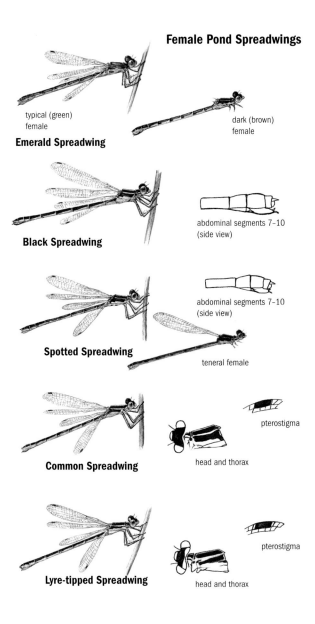

Female Pond Spreadwings

typical (green)
female

dark (brown)
female

Emerald Spreadwing

Black Spreadwing

abdominal segments 7–10
(side view)

Spotted Spreadwing

abdominal segments 7–10
(side view)

teneral female

Common Spreadwing

head and thorax

pterostigma

Lyre-tipped Spreadwing

head and thorax

pterostigma

PLATE 5   **Dancers**

Males of these two large species are extensively pruinose when mature.

Powdered Dancer *(Argia moesta)*   PAGE 53

Length: 4 cm (1.5 in.) Wing span: 4.5 to 6 cm (2 to 2.5 in.)

This is a large, pale dancer. The mature male has extensive pale gray pruinescence, appearing as if dipped in powder. The female has a mostly pale thorax that is either tan or light blue, and a darker abdomen with thin, pale rings and a pale tip.

Sooty Dancer *(Argia lugens)*   PAGE 51

Length: 4 to 5 cm (1.5 to 2 in.)
Wing span: 5.5 to 6.5 cm (2 to 2.5 in.)

The mature male has extensive slate gray or blue gray pruinescence, appearing as if dipped in soot. The tip of the abdomen is dark. The female is black, rust, and tan above, with an intricate pattern on the thorax.

PLATE 5   Dancers

Powdered Dancer

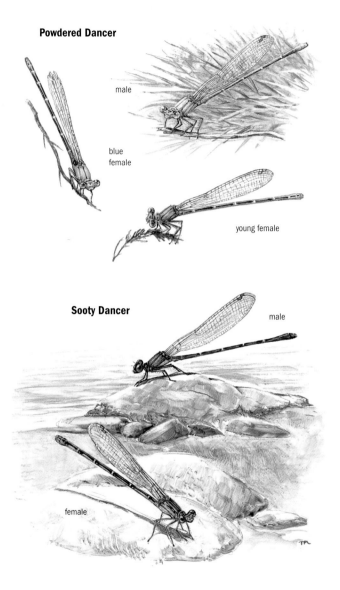

male

blue
female

young female

Sooty Dancer

male

female

PLATE 6   **Dancers**

Paiute Dancer *(Argia alberta)*   PAGE 55

Length: 3 cm (1 in.) Wing span: 4 to 4.25 cm (1.5 in.)

This small dancer's middle abdominal segments are mostly black above. The female is patterned similarly to the male; the pale areas may be tan or pale blue. The tori of the male are aligned in the same plane (in top view). This species is very similar to the Blue-ringed Dancer.

Blue-ringed Dancer *(Argia sedula)*   PAGE 54

Length: 3 to 3.5 cm (1 to 1.5 in.)
Wing span: 4 to 4.5 cm (1.5 to 2 in.)

The male is small and fairly dark above, with two deep blue or violet stripes on the thorax and powder blue rings at the bases of the middle abdominal segments. The tori slant inward, converging in a V shape. The female is paler, with a mostly tan thorax and powder blue abdominal rings. The wings are often brown tinted (smoky).

Kiowa Dancer *(Argia immunda)*   PAGE 56

Length: 3.5 cm (1.5 in.) Wing span: 4 to 5 cm (1.5 to 2 in.)

The male is pale violet or blue and black, with distinctive repeating bands of light and dark on the middle abdominal segments. The female has mottled, irregular margins of the stripes atop the thorax.

PLATE 6   Dancers

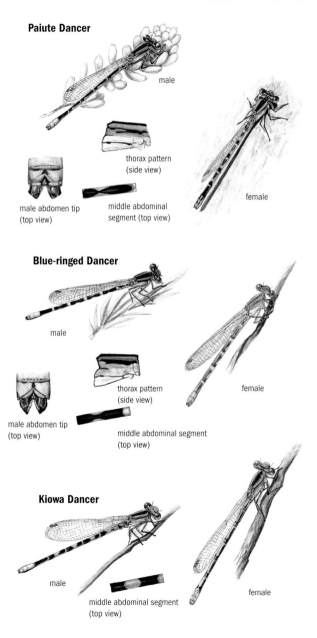

Paiute Dancer

male

thorax pattern
(side view)

male abdomen tip
(top view)

middle abdominal
segment (top view)

female

Blue-ringed Dancer

male

thorax pattern
(side view)

male abdomen tip
(top view)

middle abdominal segment
(top view)

female

Kiowa Dancer

male

middle abdominal segment
(top view)

female

PLATE 7   **Dancers**

California Dancer *(Argia agrioides)*   PAGE 58

Length: 3 to 3.5 cm (1 to 1.5 in.)
Wing span: 3.5 to 4.25 cm (1.5 in.)

The male is a blue and black dancer. The middle abdominal segments are mostly blue above. The black stripe on each side of the thorax is usually forked, but note the variation illustrated. The tori are separated by about their own width in top view. The female has pale areas on the thorax and abdomen that are blue or tan and show considerable variation in extent. It is separable from the Aztec Dancer *(A. nahuana)* only in hand (see fig. 12).

Aztec Dancer *(Argia nahuana)*   PAGE 59

Length: 3 to 3.5 cm (1 to 1.5 in.)
Wing span: 4 to 4.25 cm (1.5 in.)

This species is nearly identical to the California Dancer *(A. agrioides)*. The tori of the male are separated by a narrow gap less than the width of a torus. The female is separable from the California Dancer only in hand (see fig. 12).

Lavender Dancer *(Argia hinei)*   PAGE 57

Length: 3 to 3.5 cm (1 to 1.5 in.)
Wing span: 4 to 4.5 cm (1.5 to 2 in.)

The male is a delicate lavender and black dancer with the lower sides of the thorax white. The upper tip of each paraproct is distinctly double "toothed" in side view. The female has pale areas brown and is similar to other dancer females (see fig. 12).

PLATE 7 Dancers

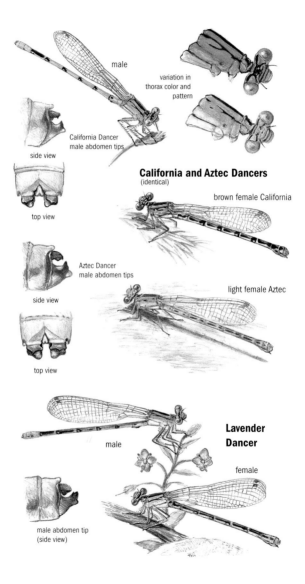

male

variation in thorax color and pattern

California Dancer male abdomen tips

side view

top view

California and Aztec Dancers
(identical)

brown female California

Aztec Dancer male abdomen tips

side view

light female Aztec

top view

Lavender Dancer

male

female

male abdomen tip
(side view)

PLATE 8   **Dancers**

Vivid Dancer *(Argia vivida)*   PAGE 60

Length: 3 to 3.75 cm (1 to 1.5 in.)
Wing span: 4 to 5 cm (1.5 to 2 in.)

This is a brilliant blue or violet blue and black dancer. Both sexes have a similar pattern on the thorax: a fairly broad, black stripe on top and a narrow, black stripe on each side that is reduced to a barely visible hairline at its midpoint (only rarely forked). Broad, pale bands on the middle abdominal segments have a black, teardrop-shaped spot on the side toward the base. Immature individuals have the pale areas nearly white to lavender or tan, and the female may be tan (gynomorphic) or blue (andromorphic) and black.

Emma's Dancer *(Argia emma)*   PAGE 61

Length: 3.5 to 4 cm (1.5 in.) Wing span: 4.5 to 5 cm (2 in.)

The thin, black central stripe atop the thorax is distinctive of both sexes. The pale areas of the mature male are purple, while the pale areas of female are blue or ochre tan. The immature male is like the tan female. The black stripes on the side of the thorax are unforked like those of the Vivid Dancer *(A. vivida)*.

PLATE 8   Dancers

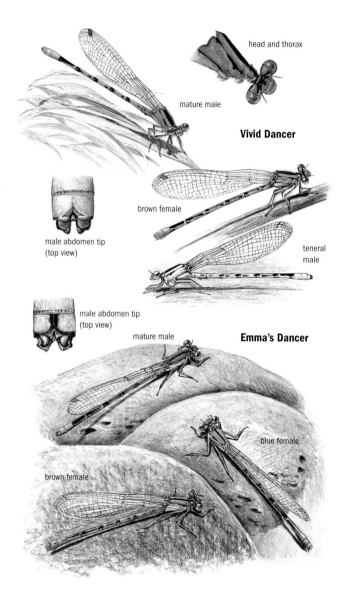

head and thorax

mature male

Vivid Dancer

brown female

male abdomen tip
(top view)

teneral
male

male abdomen tip
(top view)

mature male

Emma's Dancer

blue female

brown female

PLATE 9   Bluets

A group of look-alike species, many of them widespread and common. Males have blue and black stripes on the thorax and blue and black bands on the abdomen. Females have darker abdomens and the pale areas of the head and thorax are blue, tan, olive, or nearly white.

Taiga Bluet (*Coenagrion resolutum*) PAGE 63

Length: 2.75 to 3 cm (1 in.) Wing span: 3.5 cm (1.5 in.)

This is a small and delicate damselfly. Note the pattern on the abdomen of the male: black, U-shaped mark atop segment 2; segments 3 through 5 half blue and half black; segments 6 and 7 mostly black above. The blue stripes atop the thorax may be interrupted to form "exclamation marks." The female is similar to female American bluets, but lacks a vulvar spine (see fig. 12).

Northern Bluet
(*Enallagma cyathigerum*)   PAGE 65

Length: 3 to 4 cm (1 to 1.5 in.)
Wing span: 4 to 4.5 cm (1.5 to 2 in.)

The middle abdominal segments are mostly blue above. The male is only safely told from similar bluet species (Boreal, Familiar [*E. boreale, E. civile*]) in hand by examining abdominal appendages. Note the small upturned prong at the tips of the cerci in side view. The female cannot be confidently identified in the field.

Boreal Bluet (*Enallagma boreale*) PAGE 66

Length: 3 to 4 cm (1 to 1.5 in.)
Wing span: 4 to 4.5 cm (1.5 to 2 in.)

This species is identical to the Northern Bluet (*E. cyathigerum*) in the field. Viewed in hand under magnification, the rounded cerci of the male Boreal Bluet lack the upturned prong visible (in side view) at the tips of the cerci of the male Northern Bluet.

Alkali Bluet (*Enallagma clausum*) PAGE 67

Length: 3.5 cm (1.5 in.) Wing span: 4.5 cm (2 in.)

The male of this large bluet can be identified in hand by examining abdominal appendages: The cerci are fairly short and wedge shaped in side view. The female has abdominal segment 8 completely pale and has small pits atop the prothorax, visible with high magnification (see fig. 12).

PLATE 9   Bluets

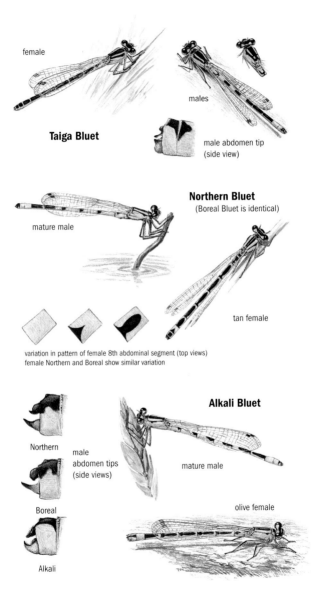

female

males

Taiga Bluet

male abdomen tip
(side view)

Northern Bluet
(Boreal Bluet is identical)

mature male

tan female

variation in pattern of female 8th abdominal segment (top views)
female Northern and Boreal show similar variation

Northern

male
abdomen tips
(side views)

Boreal

Alkali

Alkali Bluet

mature male

olive female

PLATE 10   **Bluets**

Familiar Bluet *(Enallagma civile)*   PAGE 71

Length: 3 to 4 cm (1 to 1.5 in.)
Wing span: 4 to 4.25 cm (1.5 in.)

The male typically is mostly pale on the middle abdominal segments and has a pale lobe on the trailing edge of the large, fin-shaped cerci. The male's color varies with age and temperature. The female is similar to other bluets.

Tule Bluet
(Enallagma carunculatum)   PAGE 70

Length: 2.75 to 3.75 cm (1 to 1.5 in.)
Wing span: 3 to 4 cm (1 to 1.5 in.)

The male has middle abdominal segments mostly black above. The male's cerci are tipped with a white lobe (caruncle). The female is similar to some other bluets.

Arroyo Bluet *(Enallagma praevarum)* PAGE 69

Length: 3 to 3.5 cm (1 to 1.5 in.)
Wing span: 3.5 to 4 cm (1.5 in.)

The amount of blue on the middle abdominal segments (3 through 7) of males progressively decreases from base rearward. The male cerci are two pronged, the upper prong longer than the lower one but not greatly exceeding the length of the paraproct. The female is similar to other bluets.

River Bluet *(Enallagma anna)*   PAGE 68

Length: 3 to 3.5 cm (1 to 1.5 in.)
Wing span: 4 to 4.75 cm (1.5 to 2 in.)

The male has a variable amount of blue on the middle abdominal segments. The long upper prong of the two-pronged cerci is readily visible in the hand (and occasionally in the field at close range). The female is similar to other bluets.

Double-striped Bluet
(Enallagma basidens)   PAGE 73

Length: 2 to 2.75 cm (1 in.) Wing span: 2.5 to 3 cm (1 in.)

Its tiny size and the paired, pale stripes on either side of the thorax are distinctive of both sexes.

PLATE 10   Bluets

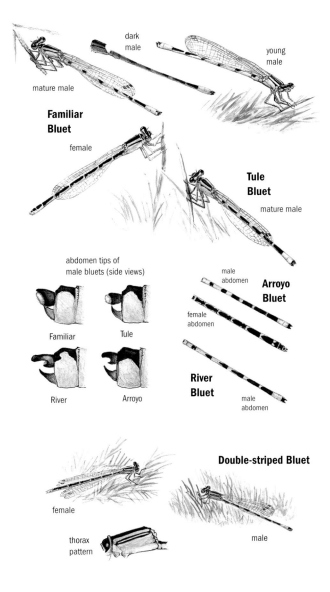

dark male

young male

mature male

Familiar Bluet

female

Tule Bluet

mature male

abdomen tips of male bluets (side views)

Familiar

Tule

River

Arroyo

male abdomen

Arroyo Bluet

female abdomen

River Bluet

male abdomen

Double-striped Bluet

female

thorax pattern

male

PLATE 11   Exclamation Damsel and Forktails

Exclamation Damsel
(*Zoniagrion exclamationis*)   PAGE 74

Length: 3 to 3.5 cm (1 to 1.5 in.)
Wing span: 4 to 4.5 cm (1.5 to 2 in.)

The male has a distinctive pattern of paired blue "exclamation marks" on the black top of the thorax and a mostly black abdomen with a large, blue patch on segments 7 through 9. The female is similar, but the segments of each exclamation mark are often fused together and the blue patch on the abdomen is on segments 7 and 8 only.

Swift Forktail (*Ischnura erratica*)   PAGE 75

Length: 3 to 3.5 cm (1 to 1.5 in.)
Wing span: 3.5 to 4 cm (1.5 in.)

This is a large forktail with two blue stripes atop the black thorax and a large, blue patch on abdominal segments 8 and 9 (sometimes extending on to segments 7 and 10 as well). The male's paraprocts have long, hook-tipped prongs. Both andromorphic (not shown) and gynomorphic females are known. Gynomorphic females are green and black and somewhat resemble female bluets, but lack a distinct vulvar spine. The teneral female is orange and black, larger than the Western Forktail (*I. perparva*).

Western Forktail
(*Ischnura perparva*)   PAGE 77

Length: 2.5 to 3 cm (1 in.) Wing span: 2.5 to 3 cm (1 in.)

The male has two narrow, green to aqua stripes atop the black thorax with green (occasionally aqua) on the sides, and a blue patch on the tip (segments 8 and 9) of the black abdomen. The male's paraprocts are distinctly forked in side view. The mature female is dark and nondescript, pruinose blue gray. The teneral gynomorphic female is orange and black (compare pattern to other forktails). Andromorphic females are very rare.

PLATE 11   Exclamation Damsel and Forktails

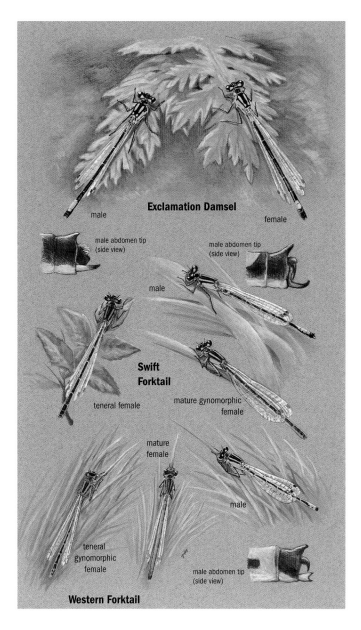

Exclamation Damsel

male

female

male abdomen tip
(side view)

male abdomen tip
(side view)

male

**Swift
Forktail**

teneral female

mature gynomorphic
female

mature
female

male

teneral
gynomorphic
female

male abdomen tip
(side view)

Western Forktail

PLATE 12   Forktails

Pacific Forktail *(Ischnura cervula)*   PAGE 81

Length: 2.5 to 3 cm (1 in.)
Wing span: 2.5 to 4 cm (1 to 1.5 in.)

The male has four blue dots atop the black thorax and a blue patch on the abdomen tip (segments 8 and 9). The andromorphic female is similar to the male, but blue atop segment 8 only. The immature gynomorphic female has a black and whitish striped thorax, two orange spots on the back of the head, and blue atop abdominal segment 8. The mature gynomorphic female is mostly slate colored above, greenish below.

San Francisco Forktail
(Ischnura gemina)   PAGE 83

Length: 2.5 to 2.75 cm (1 in.)
Wing span: 2.5 to 3.25 cm (1 to 1.5 in.)

Male a virtual twin of the Black-fronted Forktail *(I. denticollis),* with the thorax black above and turquoise on sides, and a blue patch atop abdominal segments 8 and 9. The mature gynomorphic female is mostly dark above with pale (gray, green, or tan) stripes atop the thorax. The teneral female resembles teneral female Black-fronted Forktails. The male's paraproct lacks the hooked prongs of the male Black-fronted Forktail, and the female's prothorax lacks the tiny horns present on the female Black-fronted Forktail's prothorax. It is restricted to the San Francisco Bay Area.

Black-fronted Forktail
(Ischnura denticollis)   PAGE 82

Length: 2 to 2.5 cm (1 in.) Wing span: 2.5 to 3 cm (1 in.)

The male has a black top of the thorax with green or turquoise sides and blue spots atop abdominal segments 8 and 9. The andromorphic female is rare; the gynomorphic female is dark above with green sides and pale gray, green, or tan stripes atop the thorax. The teneral female has delicate, coral pink sides to its thorax and light blue or green spots atop abdominal segments 8 and 9. This species is identical in the field to the San Francisco Forktail, but in hand note the small hook at the tips of the male's paraprocts and the pair of hornlike projections atop the female's prothorax.

PLATE 12   Forktails

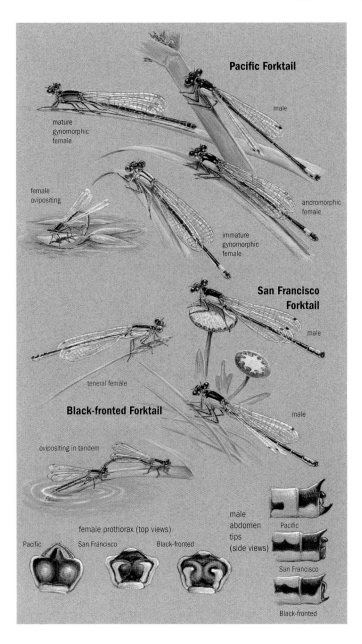

Pacific Forktail

male

mature gynomorphic female

female ovipositing

immature gynomorphic female

andromorphic female

San Francisco Forktail

male

teneral female

Black-fronted Forktail

male

ovipositing in tandem

female prothorax (top views)

Pacific   San Francisco   Black-fronted

male abdomen tips (side views)

Pacific

San Francisco

Black-fronted

PLATE 13 **Forktails**

Rambur's Forktail
(Ischnura ramburii) PAGE 78

Length: 2.75 to 3.5 cm (1 to 1.5 in.)
Wing span: 3 to 4 cm (1 to 1.5 in.)

The male and the andromorphic female have the top of
the thorax black with a pair of thin, golden green or aqua stripes, the sides
green or aqua fading to yellow green below. The top of abdominal segment 8 is blue (as is segment 9 on some individuals), and the black patch
atop each middle abdominal segment is more or less even sided. Young
gynomorphic female has a bright orange thorax with a black top stripe,
but the pale areas fade to tan or olive with age. The tips of the male's paraprocts are not upturned.

Desert Forktail *(Ischnura barberi)* PAGE 79

Length: 2.75 to 3.5 cm (1 to 1.5 in.)
Wing span: 3 to 4 cm (1 to 1.5 in.)

This species is similar to Rambur's Forktail (*I. ramburii*),
but the areas of pale coloration are somewhat more extensive and pastel hued. The male has the top of the thorax black with two
yellow green to aqua lateral stripes nearly half as wide as the black stripe
between them. Both sexes have distinctly dart-shaped, black patches atop
the middle abdominal segments. The male's paraprocts are upturned at
their tips. Andromorphic female is like male.

Citrine Forktail *(Ischnura hastata)* PAGE 84

Length: 2 to 2.5 cm (1 in.) Wing span: 2 to 3 cm (1 in.)

This is a very tiny damselfly. The distinctive male has a
mostly yellow abdomen and the pterostigma in the fore
wing not touching the costa. The teneral female is orange
and black (compare with the Desert Firetail *[Telebasis salva]*), becomes
duller with age, and is similar to the mature females of some other
species, such as Pacific and Black-fronted (*I. cervula* and *I. denticollis*)
forktails, but those are usually somewhat larger and stockier.

PLATE 13 Forktails

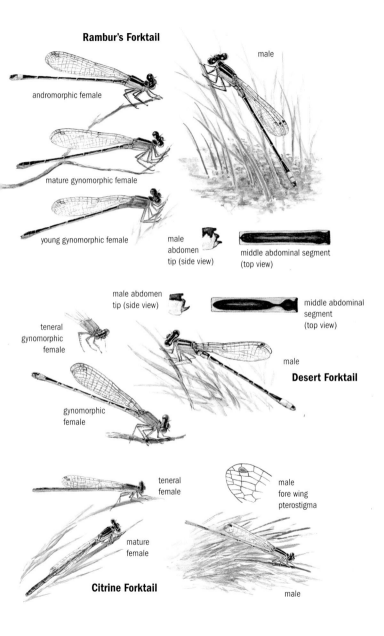

Rambur's Forktail

andromorphic female

male

mature gynomorphic female

young gynomorphic female

male abdomen tip (side view)

middle abdominal segment (top view)

male abdomen tip (side view)

middle abdominal segment (top view)

teneral gynomorphic female

male

Desert Forktail

gynomorphic female

teneral female

male fore wing pterostigma

mature female

Citrine Forktail

male

PLATE 14 **Miscellaneous Damselflies**

Desert Firetail *(Telebasis salva)*   PAGE 87

Length: 2.5 to 3 cm (1 in.) Wing span: 2.5 to 3 cm (1 in.)

This is a small, delicate damselfly that is mostly red (males) or tan (females). The long, slender abdomen projects well beyond the tip of the folded wings. It frequents floating mats of algae.

Western Red Damsel
(Amphiagrion abbreviatum)   PAGE 88

Length: 2.5 to 3 cm (1 in.)
Wing span: 3 to 4 cm (1 to 1.5 in.)

This is a stocky damselfly, with its abdomen tip barely extending beyond the wing tips when folded. The male has a mostly black thorax and a red abdomen. The female is usually warm tan or orange in color, but may be dark and nondescript instead. A large bump on the underside of the thorax is visible in the hand.

Sedge Sprite *(Nehalennia irene)*   PAGE 86

Length: 2.5 to 3 cm (1 in.)
Wing span: 2.75 to 3.25 cm (1 to 1.5 in.)

Its very tiny size and the metallic green top of the thorax distinguish both sexes. The abdomen is dark with a pale tip.

PLATE 14   Miscellaneous Damselflies

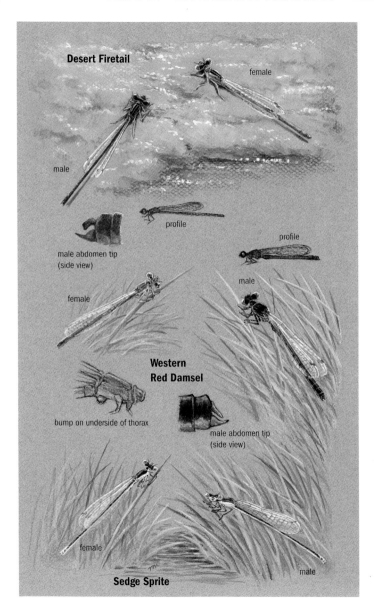

Desert Firetail

female

male

profile

male abdomen tip
(side view)

profile

female

male

**Western
Red Damsel**

bump on underside of thorax

male abdomen tip
(side view)

female

male

Sedge Sprite

PLATE 15 Green Darners

Common Green Darner
(Anax junius) PAGE 92

Length: 6.5 to 8 cm (2.5 to 3 in.)
Wing span: 9 to 10 cm (3.5 to 4 in.)

This is a large, "big-chested" dragonfly. The male has a solid green thorax and is bright blue on the base and sides of the abdomen (blue color fades when cool). The striking T-spot resembles a bull's-eye. The female also has a green thorax but is usually duller, with brown or purple on the abdomen and amber-tinted wing patches. The occipital horns of the female are visible in the hand.

Giant Darner *(Anax walsinghami)* PAGE 94

Length: males, 10 to 12 cm (4 to 4.5 in.); females, 9 to 10 cm (3.5 to 4 in.)
Wing span: 11 to 12 cm (4.5 in.)

This is our largest dragonfly. Its color pattern is similar to that of the Common Green Darner *(A. junius)*, but the much longer abdomen of the male is carried with a noticeable arc in flight. The female is similar to the female Common Green Darner, but is larger and has no occipital horns or amber tint in the wings.

PLATE 15 Green Darners

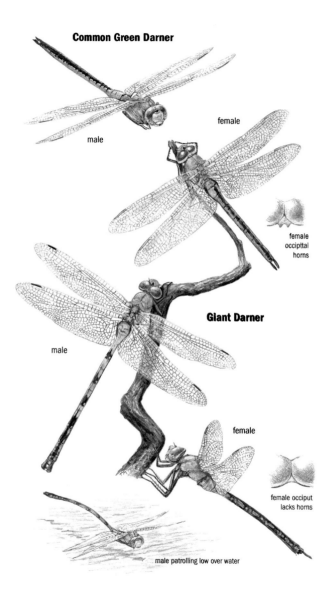

Common Green Darner

male

female

female occipttal horns

Giant Darner

male

female

female occiput lacks horns

male patrolling low over water

PLATE 16 Mosaic Darners

The two species on this plate are distinguished in the hand from all our other mosaic darners by a small tubercle on the underside of abdominal segment 1 (on both sexes).

Blue-eyed Darner
(Aeshna multicolor)
PAGE 96

Length: 6.25 to 7 cm (2.5 to 3 in.)
Wing span: 8.5 to 10 cm (3.5 to 4 in.)

The mature male has bright blue eyes, a blue face without a dark cross stripe, and blue stripes on the thorax. The female is highly variable in color but often has noticeable pale stripes on the front of the thorax and amber-tinted wings between the nodus and pterostigma (see also pl. 19).

California Darner
(Aeshna californica)
PAGE 97

Length: 5.5 to 6 cm (2 to 2.5 in.)
Wing span: 8 to 8.5 cm (3 to 3.5 in.)

This is our smallest darner. Its flight profile is less robust than that of other mosaic darners. The male has blue eyes, pale bluish white stripes on the thorax, and a pale face crossed by a black line separating the frons and clypeus. The female is similar to the Blue-eyed Darner, but lacks noticeable pale stripes on the front of the thorax and has untinted wings (see also pl. 19).

PLATE 16   Mosaic Darners

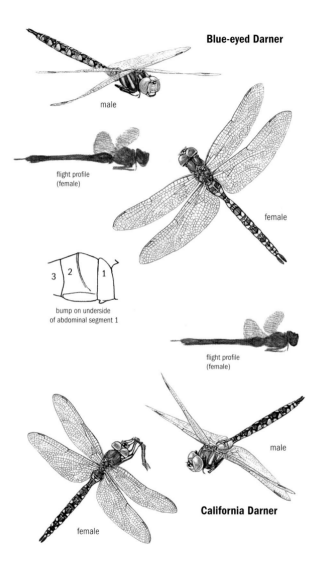

Blue-eyed Darner

male

flight profile
(female)

female

bump on underside
of abdominal segment 1

3 2 1

flight profile
(female)

male

California Darner

female

PLATE 17 Mosaic Darners and Riffle Darner

Variable Darner *(Aeshna interrupta)* PAGE 98

Length: 6.5 to 7 cm (2.5 to 3 in.) Wing span: 9 cm (3.5 in.)

The male is a relatively dark darner with thin, pale stripes on the side of the thorax, these occasionally interrupted to form paired sets of pale dashes. The pale stripes on the front of the thorax are very small or nonexistent. There is a black stripe across the face in both sexes. The female may be more or less like the male in color or brown with pale areas yellow or green. It is similar to other female darners (see also pl. 19).

Canada Darner
(Aeshna canadensis) PAGE 99

Length: 6 to 7 cm (2.5 to 3 in.)
Wing span: 8.5 to 9 cm (3.5 in.)

The anterior, pale stripe on the side of the thorax is strongly notched on both sexes, forming a vaguely shoelike shape. The face (pale blue green on male) lacks a dark cross stripe. The middle abdominal segments of both sexes have paired, blue patches at their bases on the underside. Otherwise, the female is similar in color to other mosaic darners (see also pl. 19).

Riffle Darner
(Oplonaeschna armata) PAGE 104

Length: 6.5 to 7.5 cm (2.5 to 3 in.)
Wing span: 9 to 11 cm (3.5 to 4.5 in.)

The Riffle Darner resembles a mosaic darner in overall color and pattern. The anterior, pale stripe on the side of the thorax is strongly constricted at midpoint, like that of the Canada Darner *(Aeshna canadensis)*. The male has a thin, rearward-projecting lobe at the tip of segment 10 and long, paddlelike cerci with a small but distinctive, toothed dorsal projection near the tip. The wings are clear with a short, dark pterostigma, and the radial sector does not branch in the outer wing. The rear of the head is tan.

PLATE 17   Mosaic Darners and Riffle Darner

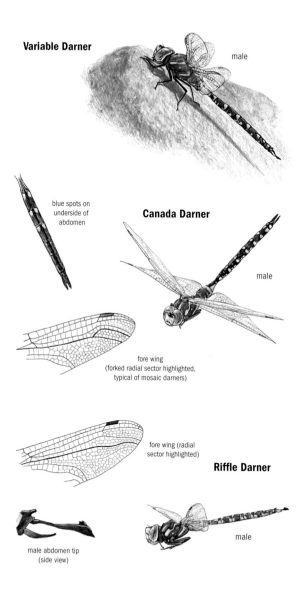

Variable Darner

male

blue spots on underside of abdomen

Canada Darner

male

fore wing
(forked radial sector highlighted,
typical of mosaic darners)

fore wing (radial
sector highlighted)

Riffle Darner

male abdomen tip
(side view)

male

PLATE 18   Mosaic Darners

The darners on this plate have male cerci that are paddle shaped and spine tipped. The female cerci often break off during oviposition (especially on the Shadow Darner *[A. umbrosa],* as shown here) and the wings of females may be tinted brown.

Shadow Darner *(Aeshna umbrosa)* PAGE 100

Length: 6.5 to 7.75 cm (2.5 to 3 in.)
Wing span: 8.5 to 10 cm (3.5 to 4 in.)

This species is difficult to distinguish from other mosaic darners except in the hand. Key features of both sexes are the lack of a dark stripe across the face, tan areas on the rear of the head behind the eyes (this area is solid black on our other mosaic darners), and paired blue (or lavender, in females) basal spots on the undersides of the middle abdominal segments. The pterostigma is typically golden tan. The female may be colored like the male or have the pale areas yellow green, like the one shown here (see also pl. 19).

Paddle-tailed Darner
(Aeshna palmata)   PAGE 101

Length: 6.5 to 7.5 cm (2.5 to 3 in.)
Wing span: 8 to 10 cm (3 to 4 in.)

This is a relatively bright mosaic darner, with the thoracic stripes typically broad and yellow or yellow green and large, blue spots on the abdomen. There is a black stripe across the yellow green face. The pterostigma is brown (black on some males). The female resembles other mosaic darners (see also pl. 19).

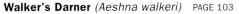

Walker's Darner *(Aeshna walkeri)* PAGE 103

Length: 7 to 7.5 cm (3 in.)
Wing span: 8.5 to 10 cm (3.5 to 4 in.)

The pale areas (especially the face and thoracic stripes) are much less yellow in tone than our other mosaic darners with paddlelike male appendages. The face is usually bone white with a slight blue green tint and a relatively faint, brown cross stripe. The stripes on the side of the thorax are bluish white. The pterostigma is black, including on females (see also pl. 19).

PLATE 18 Mosaic Darners

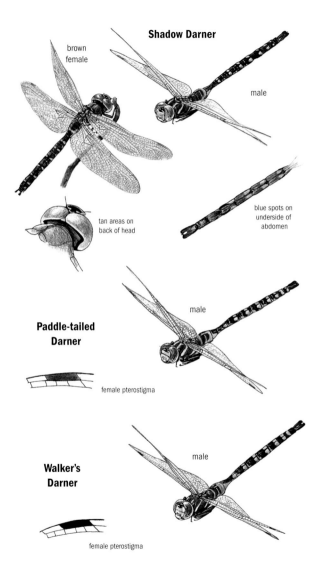

Shadow Darner

brown female

male

tan areas on back of head

blue spots on underside of abdomen

Paddle-tailed Darner

male

female pterostigma

Walker's Darner

male

female pterostigma

PLATE 19   **Mosaic Darners**

Key features to note on perched mosaic darners in the field: (1) the face pattern, including the T-spot; (2) the stripes on the thorax; and (3) the abdomen tip of the male. *See additional field marks, range maps, and length and wing span information on plates 16, 17, and 18.*

Blue-eyed Darner *(Aeshna multicolor)*   PAGE 96

Face without a dark stripe, T-spot with thin, straight crossbar. Stripes on side of thorax relatively broad and straight edged. Male's cerci are two pronged at tip in side view.

California Darner *(Aeshna californica)*   PAGE 97

The face has a dark stripe and a thick T-spot with an angled crossbar. The anterior stripe on the side of the thorax tapers to a point at the upper end. The male's cerci are of simple shape.

Variable Darner *(Aeshna interrupta)*   PAGE 98

The face has a dark stripe, and the T-spot has an angled crossbar. The thin stripes on the side of the thorax may be interrupted to form paired dash marks (as shown here). The male's cerci are simple and parallel sided.

Canada Darner *(Aeshna canadensis)*   PAGE 99

The face lacks a cross stripe. The anterior stripe on the side of the thorax is strongly notched in the middle. The male's cerci are of simple shape.

Shadow Darner *(Aeshna umbrosa)*   PAGE 100

The face is without a dark stripe, the T-spot with a short stem and straight-edged crossbar. The stripes on the side of the thorax are of even width and straight sided. The male's cerci are paddle shaped with a long spine, and the top of abdominal segment 10 is without pale spots.

Paddle-tailed Darner *(Aeshna palmata)*   PAGE 101

The face has a black cross stripe and a T-spot with a relatively short, thin stem and a straight-edged crossbar. The stripes on the thorax are relatively broad and usually yellow green. The male's cerci are paddle shaped with a long spine, and there are blue spots atop segment 10.

Walker's Darner *(Aeshna walkeri)*   PAGE 103

The face has a brown cross stripe and a T-spot with a relatively long, thick stem and an angled crossbar. The stripes on the thorax are pale blue to white. The male's cerci are paddle shaped with a short spine, and the top of abdominal segment 10 is all dark without pale spots.

PLATE 19   Mosaic Darners

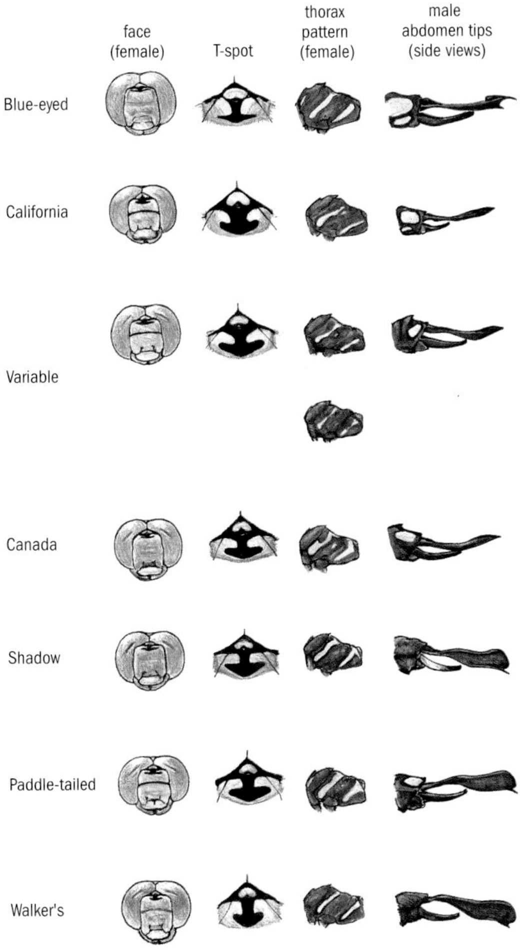

Mosaic Darners

	face (female)	T-spot	thorax pattern (female)	male abdomen tips (side views)
Blue-eyed				
California				
Variable				
Canada				
Shadow				
Paddle-tailed				
Walker's				

PLATE 20   Petaltail and Clubtails

Black Petaltail *(Tanypteryx hageni)*   PAGE 90

Length: 5.5 cm (2 in.) Wing span: 7.5 cm (3 in.)

This is a large, unwary, black-and-yellow dragonfly of spring-fed bogs. The dark compound eyes don't come in contact. The pterostigma is long and extremely narrow. The female's abdomen is blunt tipped, and the male's cerci are leaflike.

Pacific Clubtail *(Gomphus kurilis)*   PAGE 107

Length: 5.25 cm (2 in.) Wing span: 6 to 7 cm (2.5 to 3 in.)

This clubtail is black with pale markings of green and yellow, and blue gray eyes. A key feature is the black, diagonal slash separating the two pale yellow or gray green patches on the side of the thorax. It often flushes with a distinctive, bounding flight.

Grappletail
(Octogomphus specularis)   PAGE 106

Length: 5.25 cm (2 in.) Wing span: 6.5 cm (2.5 in.)

Both sexes have a unique pattern on the thorax: a pale patch shaped like a goblet or urn on the top (front), separated from the pale sides by broad, black stripes. The pale areas of the thorax may be yellow, green, or gray. Abdominal segment 10 of the male forms an expanded club. The top of the abdomen is mostly black, unlike other clubtails.

PLATE 20　Petaltail and Clubtails

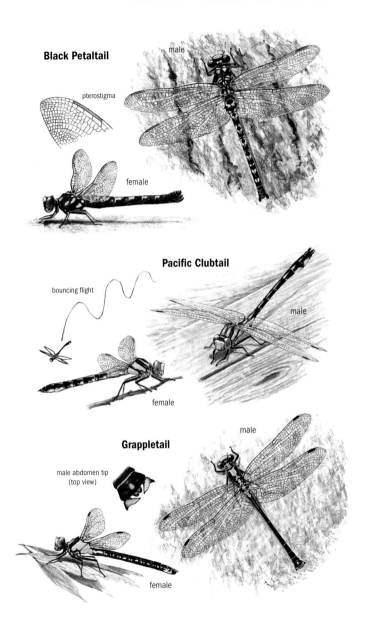

Black Petaltail

pterostigma

male

female

Pacific Clubtail

bouncing flight

male

female

Grappletail

male abdomen tip
(top view)

male

female

PLATE 21 **Hanging Clubtails**

Unlike other clubtails, these species typically hang vertically from bushes and trees.

Olive Clubtail *(Stylurus olivaceus)* PAGE 109

Length: 5.5 to 6 cm (2 to 2.5 in.) Wing span: 8 cm (3 in.)

This is a rather drab olive gray and black dragonfly. The front of the thorax is dark brown or black with a pale olive gray, horseshoe-shaped mark. The sides of the thorax are unmarked, pale olive gray.

Russet-tipped Clubtail
(Stylurus plagiatus) PAGE 110

Length: 5.5 to 6.5 cm (2 to 2.5 in.)
Wing span: 7 to 8 cm (3 in.)

The thorax is green striped with dark brown on the front and sides. The abdomen tip is marked with yellow orange. The eyes are turquoise.

Brimstone Clubtail
(Stylurus intricatus) PAGE 111

Length: 4 to 5.5 cm (1.5 to 2 in.)
Wing span: 5.5 to 6.5 cm (2 to 2.5 in.)

This small and pale yellow clubtail has reduced dark markings on the thorax and abdomen compared to our other hanging clubtails. The abdomen is mostly yellow, paler and green tinged on segments 1 through 6, brighter on segments 7 through 10.

PLATE 21   Hanging Clubtails

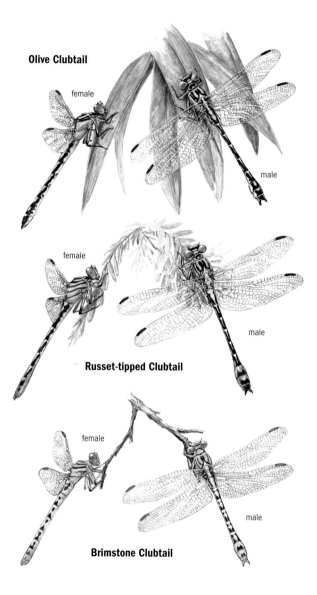

Olive Clubtail

female

male

Russet-tipped Clubtail

female

male

Brimstone Clubtail

female

male

PLATE 22   **Snaketails**

Green thorax with brown stripes, and a snakeskin-like pattern of yellow and black markings on the abdomen.

Bison Snaketail
(Ophiogomphus bison)   PAGE 116

Length: 5 cm (2 in.) Wing span: 6 to 6.5 cm (2.5 in.)

The thorax has broad, green and brown stripes in front with relatively straight margins. The male's cerci and epiproct are relatively long and straight. The female's occipital horns are close together at the top edge of the occiput.

Sinuous Snaketail
(Ophiogomphus occidentis)   PAGE 115

Length: 5 cm (2 in.) Wing span: 6 to 6.5 cm (2.5 in.)

The brown stripe on either upper side of the thorax has a pale thin, wavy line running down its center. Central Valley populations are paler with the dark markings less extensive. The male's epiproct is sharply curved upward. The female has two forward-projecting occipital horns and two hooked postoccipital horns.

Great Basin Snaketail
(Ophiogomphus morrisoni)   PAGE 113

Length: 5 to 5.25 cm (2 in.) Wing span: 6.5 cm (2.5 in.)

The pale green stripe in the brown stripe on either side of the thorax above is relatively straight, not wavy. The eyes are aqua blue. Sierran populations have more extensive dark markings than do Great Basin populations. The male's cerci are short and stout, not longer than the epiproct. The female's occiput is usually without horns but may have a widely spaced pair.

Pale Snaketail
(Ophiogomphus severus)   PAGE 112

Length: 5 cm (2 in.) Wing span: 6 to 6.5 cm (2.5 in.)

The front of the thorax is mostly yellow green with a thin brown, center stripe and brown spindle-shaped marks in the upper corners. It is paler overall than other snaketails, although some female Great Basin Snaketails *(O. morrisoni)* are similar and may be indistinguishable. The male's cerci are stout and about the same length or slightly longer than the epiproct. The female's occiput lacks horns.

PLATE 22   Snaketails

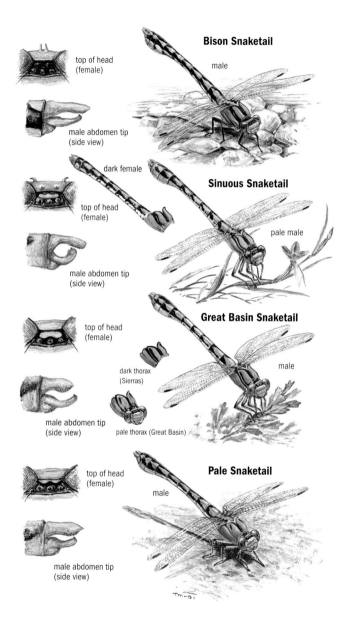

Bison Snaketail

male

top of head (female)

male abdomen tip (side view)

dark female

Sinuous Snaketail

top of head (female)

pale male

male abdomen tip (side view)

Great Basin Snaketail

top of head (female)

dark thorax (Sierras)

male

pale thorax (Great Basin)

male abdomen tip (side view)

top of head (female)

Pale Snaketail

male

male abdomen tip (side view)

PLATE 23   **Ringtails and Sanddragon**

White-belted Ringtail
(Erpetogomphus compositus)   PAGE 117

Length: 4.5 to 5.5 cm (2 in.) Wing span: 7 cm (3 in.)

It looks like a composite of different species: eyes blue gray; thorax striped with brown, green, and white; abdomen mostly black with white rings; and a bright orange yellow club (on males). The male's cerci taper to a smooth, blunt tip. The female's occiput has a wavy ridge along the top.

Serpent Ringtail
(Erpetogomphus lampropeltis)   PAGE 118

Length: 4 to 5.5 cm (1.5 to 2 in.) Wing span: 7 cm (3 in.)

This ringtail is similar to, but more drab than, the White-belted Ringtail (E. compositus), with reduced or broken pale green stripes on the brown thorax and a darker abdomen tip. The male's cerci are pinched to a narrow point at the tip. The female's occiput has a dark rear margin.

Gray Sanddragon
(Progomphus borealis)   PAGE 119

Length: 5.5 cm (2 in.) Wing span: 7 cm (3 in.)

The thorax is brown in front with elaborate yellow markings. The sides of the thorax are gray with a brown diagonal stripe. The slender abdomen is mostly black with narrow, creamy yellow triangles and is tipped on males with long white or pale yellow cerci. The male epiproct, which is black and divided in two, is unique among our dragonflies.

PLATE 23   Ringtails and Sanddragon

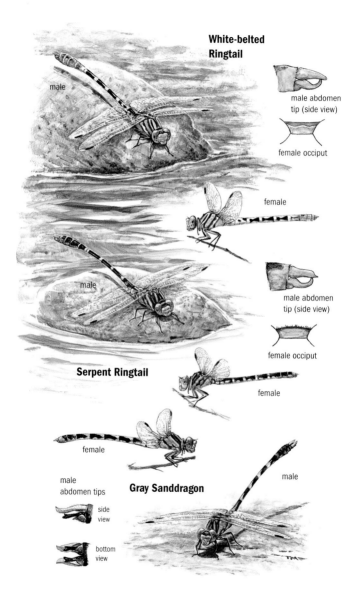

White-belted Ringtail

male

male abdomen tip (side view)

female occiput

female

male abdomen tip (side view)

female occiput

Serpent Ringtail

male

female

female

Gray Sanddragon

male abdomen tips

side view

bottom view

male

PLATE 24   Spiketail and River Cruiser

Pacific Spiketail
(Cordulegaster dorsalis)   PAGE 121

Length: 7 to 8.5 cm (3 to 3.5 in.)
Wing span: 9 to 10.5 cm (3.5 to 4 in.)

    This is a large and impressive black-and-yellow dragonfly. The thorax has two broad yellow stripes in front and two on either side. The eyes of the adults are brilliant aqua blue. The female has a long, spike-like ovipositor. Populations east of the Sierra Nevada have pale areas more extensive.

Western River Cruiser
(Macromia magnifica)   PAGE 123

Length: 7 to 7.5 cm (3 in.)
Wing span: 9 to 10 cm (3.5 to 4 in.)

    This is a large, brown and yellow dragonfly. There is only a single, pale yellow stripe on the side of the thorax. The eyes are gray green. There is a prominent T-spot on the frons. Both the adult and larva have extremely long legs. The male's abdomen is typically arched in flight.

PLATE 24 Spiketail and River Cruiser

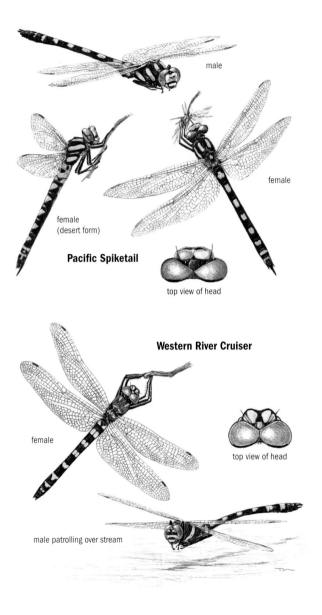

male

female

female
(desert form)

Pacific Spiketail

top view of head

Western River Cruiser

female

top view of head

male patrolling over stream

PLATE 25 Emeralds

Dark, mostly black dragonflies with brilliant emerald green eyes on mature individuals. They are found at higher elevations for the most part in California, from the southern Sierra Nevada north.

American Emerald
(Cordulia shurtleffii)

PAGE 125

Length: 4.5 to 5 cm (2 in.) Wing span: 6.5 cm (2.5 in.)

The thorax is iridescent green and bronze and without pale marks. The male's cerci are relatively short and simple in shape. There is a white ring at the base of abdominal segment 3. The female's cerci are relatively short. Males patrol over lakes and ponds.

Mountain Emerald
(Somatochlora semicircularis)

PAGE 126

Length: 5 to 5.5 cm (2 in.)
Wing span: 6.5 to 7 cm (2.5 to 3 in.)

The thorax has two yellow spots on the sides. The male's cerci are long and curved in top view. The female's cerci are relatively long. Males patrol over bogs and wet meadows.

Ringed Emerald
(Somatochlora albicincta)

PAGE 127

Length: 4.5 to 5 cm (2 in.) Wing span: 6.5 cm (2.5 in.)

The bases of the abdominal segments have broken white rings. There is a single, pale creamy white spot on the side of the thorax. The male's cerci are relatively straight and long with a sharp hook at the tip.

PLATE 25 Emeralds

young female

female abdomen tip (side view)

male

male abdomen tip (side view)

American Emerald

female abdomen tip (side view)

young female

male

male abdomen tip (side view)

Mountain Emerald

young female

male

Ringed Emerald

male abdomen tip (side view)

male American Emerald patrolling over water

male Mountain Emerald patrolling over bog

PLATE 26   **Baskettails and Skimmers**

Beaverpond Baskettail
(Tetragoneuria canis)   PAGE 128

Length: 4.5 to 5 cm (2 in.) Wing span: 6.5 cm (2.5 in.)

This is a black, brown, and butterscotch dragonfly with blue green eyes. The T-spot on the frons lacks a crossbar. The male's cerci resemble a dog's head in side view. The female's cerci are relatively short.

Spiny Baskettail
(Tetragoneuria spinigera)   PAGE 129

Length: 4.5 cm (2 in.) Wing span: 6 to 6.5 cm (2.5 in.)

This species is virtually identical to the Beaverpond Baskettail *(T. canis)* in the field. The T-spot is complete (has a crossbar). The male's cerci have a rounded tip and small downward-projecting spine at the midpoint. The female's cerci are relatively long.

Four-spotted Skimmer
(Libellula quadrimaculata)   PAGE 158

Length: 4.5 cm (2 in.) Wing span: 6.5 to 7 cm (2.5 to 3 in.)

This is a rather drab brown or olive dragonfly with yellow stripes on the sides of the black-tipped abdomen, small black patches (the four "spots") at the nodus of each wing, and a black and tan triangular patch at the base of each hind wing. Young individuals are brighter, with golden color along the leading edges of the wings.

Chalk-fronted Corporal
(Ladona julia)   PAGE 156

Length: 4 to 4.5 cm (1.5 to 2 in.)
Wing span: 6 to 7 cm (2.5 to 3 in.)

The mature male is black with mostly clear wings (some black at bases), two white bars on the front of the thorax (the "corporal's stripes"), and a white base to the abdomen. The mature female is a drab dragonfly with a very short and stubby abdomen, black at tip and grayer toward the base. The immature female (not shown) is somewhat paler and browner.

PLATE 26   Baskettails and Skimmers

heads (top views)

Spiny

Beaverpond

Beaverpond Baskettail
(Spiny Baskettail nearly identical)

male

teneral female

Spiny

Beaverpond

female cerci (side views)

Spiny

Beaverpond

male abdomen tips (side views)

flight profile

young female

mature male

**Four-spotted
Skimmer**

flight profile

mature female

male

Chalk-fronted Corporal

PLATE 27   Whitefaces

Small, dark dragonflies with white faces, found in the mountains of northern California. They are all similar in appearance, and usually require examination of male abdominal appendages (shown here) and the male hamules and female vulvar lamina (fig. 13) for sure identification.

Crimson-ringed Whiteface
(Leucorrhinia glacialis)   PAGE 132

Length: 3.5 to 4 cm (1.5 in.)
Wing span: 5.5 to 6 cm (2 to 2.5 in.)

The male has red markings on the thorax and basal abdominal segments, but is usually all black on segments 4 through 10. The male is nearly identical to the Red-waisted *(L. proxima)* male. The female has either red or yellow markings on the thorax and abdomen and is virtually identical to other female whitefaces. Both sexes have two rows of cells between the radial sector and radial planate.

Red-waisted Whiteface
(Leucorrhinia proxima)   PAGE 133

Length: 3.5 cm (1.5 in.) Wing span: 5 to 5.5 cm (2 in.)

Rare and local in the Lassen Peak region. It is identical in the field to the Crimson-ringed Whiteface *(L. glacialis)*. In the hand, note the single row of cells between the radial sector and radial planate.

Hudsonian Whiteface
(Leucorrhinia hudsonica)   PAGE 131

Length: 3 to 3.5 cm (1 to 1.5 in.)
Wing span: 5 to 6 cm (2 to 2.5 in.)

The male and some females have red spots atop middle abdominal segments. Some females and the young male are yellow where the mature male is red. There is a single row of cells between the radial sector and radial planate.

Dot-tailed Whiteface
(Leucorrhinia intacta)   PAGE 134

Length: 3.5 cm (1.5 in.) Wing span: 5.75 cm (2 in.)

The mature male is mostly black with a square, yellow "dot" atop abdominal segment 7. Young individuals and females have some yellow or reddish markings on the thorax and abdomen similar to other whitefaces, but also have the "dot" on segment 7.

PLATE 27 Whitefaces

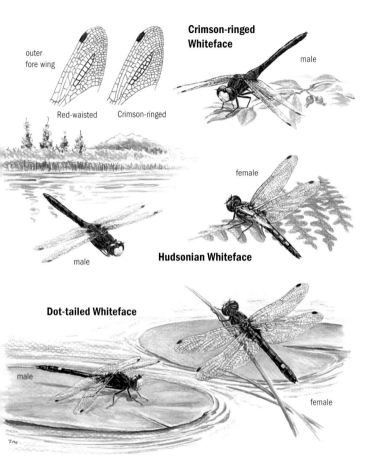

outer
fore wing

Red-waisted   Crimson-ringed

Crimson-ringed Whiteface

male

female

male

Hudsonian Whiteface

Dot-tailed Whiteface

male

female

abdomen tips of male whitefaces

dot-tailed   Hudsonian   Crimson-ringed   Red-waisted

PLATE 28 **Meadowhawks**

Western Meadowhawk
(Sympetrum occidentale)   PAGE 137

Length: 3 to 4 cm (1 to 1.5 in.)
Wing span: 4.5 to 6 cm (2 to 2.5 in.)

The best distinguishing mark is an amber brown wash on the basal half of the wings, although this is sometimes barely visible on some females and young males. The sides of the thorax are yellow or olive with black, flamelike stripes. The abdomen is deep yellow maturing to red, with a black stripe on the sides and black spots atop segments 8 and 9.

Black Meadowhawk
(Sympetrum danae)   PAGE 136

Length: 2.75 to 3 cm (1 in.) Wing span: 4.5 to 5 cm (2 in.)

This is a small black and yellow meadowhawk, with no red markings. The male is nearly all black, including the face. The side of the thorax is distinctively patterned with yellow dots in a broad, black band. The female may have yellow washes at the base and nodus of the wings.

Yellow-legged Meadowhawk
(Sympetrum vicinum)   PAGE 140

Length: 3 to 3.5 cm (1 to 1.5 in.) Wing span: 4.5 cm (2 in.)

The mature male has a red abdomen and a reddish brown thorax, and the female is yellow and tan; otherwise it is a rather plain dragonfly without black or white markings. The legs are pale. The female has swollen basal abdominal segments and a large, spout-shaped vulvar lamina (fig. 13). Older females may show red on the abdomen.

PLATE 28 Meadowhawks

Western Meadowhawk

mature female

male

immature male

thorax pattern (side view)

thorax pattern (side view)

Black Meadowhawk

male

female

male

teneral female

Yellow-legged Meadowhawk

PLATE 29   Meadowhawks

White-faced Meadowhawk
(Sympetrum obtrusum)

PAGE 142

Length: 3 to 4 cm (1 to 1.5 in.)
Wing span: 4.5 to 5.5 cm (2 in.)

The mature male has a red abdomen with black side stripes, an unmarked brown abdomen, and a white face. The female is rather nondescript tan and yellow, with black stripes on the sides of the abdomen, dark wing veins, and clear wings. Older females may be red atop the abdomen.

Cherry-faced Meadowhawk
(Sympetrum internum)

PAGE 141

Length: 3 to 3.5 cm (1 to 1.5 in.)
Wing span: 5 to 6 cm (2 to 2.5 in.)

The male is like the White-faced Meadowhawk (*S. obtrusum*), but has a red face when mature and rusty veins in the front part of the wings. The female is like the White-faced, but has tawny wing veins (and sometimes an amber wash in the wing base). In the hand, compare vulvar laminae of the two species (fig. 13).

Saffron-winged Meadowhawk
(Sympetrum costiferum)

PAGE 139

Length: 3 to 4 cm (1 to 1.5 in.)
Wing span: 5 to 6 cm (2 to 2.5 in.)

The veins in the leading edges of the wing are rusty (in the male) or amber, and the costal stripe is washed with amber on the female and young male. The pterostigma is relatively large and pale yellow, with black edging (becomes orange on the mature male). The male has a dull, red abdomen with a relatively narrow stripe of black on either side. Some older females show red on the abdomen, too.

PLATE 29 Meadowhawks

male

young female

White-faced Meadowhawk

pterostigma

Cherry-faced Meadowhawk

male

young female

Saffron-winged Meadowhawk

female

male

pterostigma

PLATE 30   **Meadowhawks**

Cardinal Meadowhawk
(Sympetrum illotum)   PAGE 147

Length: 3 to 4 cm (1 to 1.5 in.)
Wing span: 5.5 to 6 cm (2 to 2.5 in.)

The mature male is mostly red, brightest on the abdomen, with two white spots on the sides of the thorax. The wings have black streaks and a rusty wash at the base. The female has a similar pattern, but is tan where the male is red and has an amber wash in the wings.

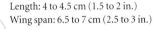

Red-veined Meadowhawk
(Sympetrum madidum)   PAGE 144

Length: 4 to 4.5 cm (1.5 to 2 in.)
Wing span: 6.5 to 7 cm (2.5 to 3 in.)

The wing veins along the leading edge are red (in the mature male) or amber (in the female and young male). The sides of the warm brown thorax have two white stripes. The female and the young male have an olive yellow abdomen with thin, black lines along the sides, while the mature male's abdomen is rusty red. There are two rows of cells between the radial sector and the radial planate.

PLATE 30 Meadowhawks

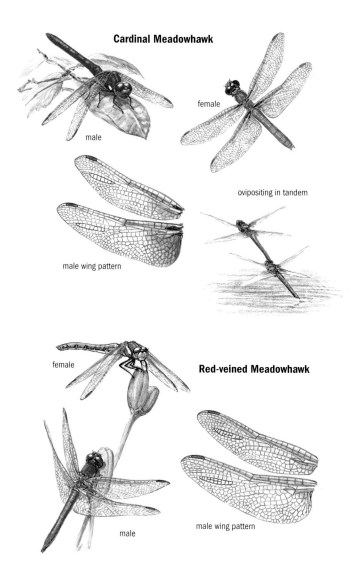

Cardinal Meadowhawk

male

female

ovipositing in tandem

male wing pattern

female

Red-veined Meadowhawk

male

male wing pattern

PLATE 31   Meadowhawks

Variegated Meadowhawk
(Sympetrum corruptum)   PAGE 145

Length: 4 to 4.5 cm (1.5 to 2 in.)
Wing span: 6 to 7.5 cm (2.5 to 3 in.)

This species is highly variable in color, ranging from pastel pink to gold and tan to dull gray, but it has distinctive white, black-rimmed spots along the side of the abdomen. The thorax has two white stripes with yellow dots in their lower ends on each side and two white stripes on the front. The mature male is often washed nearly throughout with a dull red or pink color, but the yellow dots on the thorax and the row of white spots along the abdomen are still at least faintly visible. The pterostigma is noticeably bicolored, dark in the middle and pale at both ends. There are two rows of cells between the radial sector and radial planate.

Striped Meadowhawk
(Sympetrum pallipes)   PAGE 142

Length: 3.5 to 4 cm (1.5 in.)
Wing span: 6 to 6.5 cm (2.5 in.)

The brown thorax has two white stripes in the front and two on either side. The abdomen is tan with a narrow black and white side stripe on young individuals, becoming tomato red above on mature individuals. The lower basal half of the abdomen develops a white pruinescence. The legs may be dark or pale on either sex. It has only a single row of cells between radial sector and radial planate.

PLATE 31   Meadowhawks

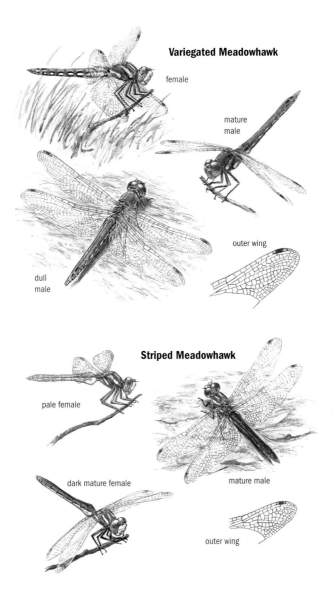

Variegated Meadowhawk

female

mature
male

outer wing

dull
male

Striped Meadowhawk

pale female

mature male

dark mature female

outer wing

PLATE 32   Amberwing, Pennant, and Skimmer

Mexican Amberwing
(Perithemis intensa)   PAGE 152

Length: 2.5 cm (1 in.) Wing span: 4 to 4.5 cm (1.5 to 2 in.)

Its small size, dumpy proportions, and large patches of yellow in the wings make this little dragonfly unlike any other in California. The male in the field appears nearly all golden yellow, including its wings.

Red-tailed Pennant
(Brachymesia furcata)   PAGE 148

Length: 4 to 4.5 cm (1.5 to 2 in.)
Wing span: 6.5 to 7.5 cm (2.5 to 3 in.)

The male has an olive drab thorax and a bright red abdomen with ski-tipped cerci. The drab brown female has a white stripe between the wings atop the thorax and abdomen base and small, black spots atop segments 8 and 9. The hind wings have a small wash of amber at their bases.

Roseate Skimmer
(Orthemis ferruginea)   PAGE 168

Length: 5.5 cm (2 in.) Wing span: 8 to 9.5 cm (3 to 3.5 in.)

The rosy pink and plum colors of the male are unlike any other dragonfly in the state. The female has a tawny orange abdomen and a brown-and-white-striped thorax. The wing has a distinctive brown stripe along the leading edge from the pterostigma to the tip.

PLATE 32   Amberwing, Pennant, and Skimmer

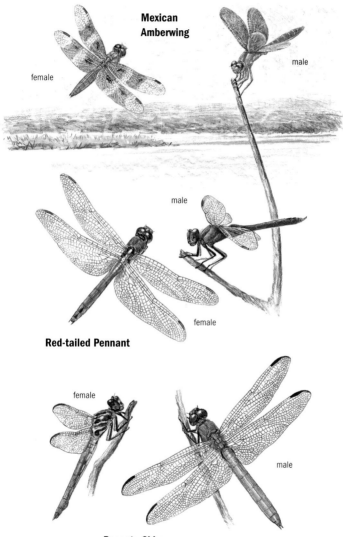

Mexican Amberwing

female

male

male

female

Red-tailed Pennant

female

male

Roseate Skimmer

PLATE 33   **Blue Dasher and Pondhawk**

Blue Dasher
(Pachydiplax longipennis)   PAGE 150

Length: 3.5 to 4 cm (1.5 in.)
Wing span: 5.5 to 6 cm (2 to 2.5 in.)

The mature male has a mostly blue, pruinose body, emerald green eyes, and a white face. The female and young male have brown eyes, a black and yellow striped thorax, and rows of yellow dashes along the middle segments of the black abdomen. The maturing male develops pruinescence on the abdomen first.

Western Pondhawk
(Erythemis collocata)   PAGE 149

Length: 4 cm (1.5 in.) Wing span: 6.5 cm (2.5 in.)

The mature male has a blue, pruinose body and green face, with dark blue green eyes. The female and young male are mostly green with variable amounts of tan and black on the abdomen. The maturing male has green on the thorax and a blue abdomen. This species often perches on the ground.

PLATE 33   Blue Dasher and Pondhawk

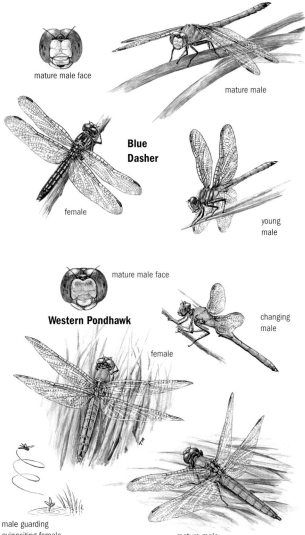

mature male face

mature male

Blue Dasher

female

young male

mature male face

Western Pondhawk

changing male

female

male guarding ovipositing female

mature male

Male skimmers with black and white wing patches. Mature males are shown. Immature males lack pruinescence on wings or body, the pattern on abdomen and thorax similar to that of females (pl. 35).

Eight-spotted Skimmer
(Libellula forensis)   PAGE 160

Length: 5 cm (2 in.) Wing span: 7.5 to 8 cm (3 in.)

The eight black wing spots include a large, black patch at each wing base and a vaguely figure eight–shaped patch behind each nodus. White pruinescence develops on the basal half of the wing around the basal black spots and out near the pterostigma. The wing tip is clear. It is pruinose pearl gray on the abdomen and the front of the thorax.

Twelve-spotted Skimmer
(Libellula pulchella)   PAGE 159

Length: 5.5 cm (2 in.) Wing span: 8.5 to 9 cm (3.5 in.)

The wings are spotted much like those of the Eight-spotted Skimmer *(L. forensis),* but with four additional black spots, one on each wing tip.

Hoary Skimmer
(Libellula nodisticta)   PAGE 161

Length: 4.5 to 5 cm (2 in.) Wing span: 7.5 to 8 cm (3 in.)

This species has eight spots as does the Eight-spotted Skimmer *(L. forensis),* but these are reduced in size, with only a small, black spot at the nodus.

Continued on the next plate

Skimmers with black
and white wings (males)

Eight-spotted Skimmer

Twelve-spotted Skimmer

Hoary Skimmer

Continued from the previous plate

Widow Skimmer
(Libellula luctuosa)   PAGE 163

Length: 4.5 to 5 cm (2 in.) Wing span: 7.5 to 8 cm (3 in.)

The basal half of each wing is washed with dark brown or black, darkest in the band behind the nodus, pruinose white from the nodus to pterostigma. The abdomen and the front of the thorax are pruinose white.

Common Whitetail
(Plathemis lydia)   PAGE 153

Length: 4.25 to 4.75 cm (1.5 to 2 in.)
Wing span: 6.5 to 7.5 cm (2.5 to 3 in.)

The broad, black band on the middle of each wing and the bright white abdomen are distinctive.

Desert Whitetail
(Plathemis subornata)   PAGE 154

Length: 4 to 5 cm (1.5 to 2 in.)
Wing span: 6.5 to 7.5 cm (2.5 to 3 in.)

This species is similar to Common Whitetail (*P. lydia*), but with extensive white pruinescence on the basal half of wing. Note also that the black wing bands have wavy borders and often appear paler centrally.

Skimmers with black
and white wings (males)

Widow Skimmer

Common Whitetail

Desert Whitetail

PLATE 35 Black and White Skimmers, Females

Female skimmers with black wing patches. *See range maps on plate 34*

Eight-spotted Skimmer *(Libellula forensis)*   PAGE 160

Length: 5 cm (2 in.) Wing span: 7.5 to 8 cm (3 in.)

There are eight large, black wing patches, one at each wing base and one behind each nodus. The mature female may develop white pruinescence on wings like the male, unlike other species on this plate.

Twelve-spotted Skimmer *(Libellula pulchella)*   PAGE 159

Length: 5.5 cm (2 in.) Wing span: 8.5 to 9 cm (3.5 in.)

The wings are spotted much like those of the Eight-spotted Skimmer *(L. forensis),* but with four additional black spots, one on each wing tip. The long, relatively slender abdomen has pale yellow side stripes.

Hoary Skimmer *(Libellula nodisticta)*   PAGE 161

Length: 4.5 to 5 cm (2 in.) Wing span: 7.5 to 8 cm (3 in.)

The female is patterned like the male, with a black patch at each wing base and a small, black spot at the nodus.

Widow Skimmer *(Libellula luctuosa)*   PAGE 163

Length: 4.5 to 5 cm (2 in.) Wing span: 7.5 to 8 cm (3 in.)

The basal half of each wing is washed with dark brown or black, darkest in the band behind the nodus. The dark patches are reduced to rather pale brown washes on teneral individuals. A small, brown wash is often present on the wing tip. The yellow stripe atop the thorax forks at the base of the abdomen to become two yellow side stripes.

Common Whitetail *(Plathemis lydia)*   PAGE 153

Length: 4.25 to 4.75 cm (1.5 to 2 in.) Wing span: 6.5 to 7.5 cm (2.5 to 3 in.)

The wing pattern is very different from that of the male (pl. 34), but very similar to the female Twelve-spotted Skimmer *(L. pulchella).* The broad, short abdomen has a row of white dashes along each side.

Desert Whitetail *(Plathemis subornata)*   PAGE 154

Length: 4 to 5 cm (1.5 to 2 in.) Wing span: 6.5 to 7.5 cm (2.5 to 3 in.)

Each wing has two wavy black or brown bands, one behind the nodus, the other behind the pterostigma. It may have a light brown wash between these two bands.

PLATE 35   Black and White Skimmers, Females

Skimmers with dark wing patches (females)

Eight-spotted Skimmer

Widow Skimmer

Twelve-spotted Skimmer

Common Whitetail

Hoary Skimmer

Desert Whitetail

PLATE 36   Pale-eyed Skimmers

Comanche Skimmer
(Libellula comanche)   PAGE 164

Length: 4.75 to 5.5 cm (2 in.)
Wing span: 7 to 9 cm (3 to 3.5 in.)

This is our only dragonfly with the pterostigma mostly white. The mature male is pruinose blue with a white face and pale eyes. Note the broad, brown and white stripes on the thorax and the brown wash at the wing tips of the female.

Bleached Skimmer
(Libellula composita)   PAGE 165

Length: 4 to 5 cm (1.5 to 2 in.)
Wing span: 7.5 to 8 cm (3 in.)

The males are very pale overall, with a washed out or bleached appearance. White costa of both sexes is unique. Note the brown and amber patches at the wing bases and the pale eyes. It often has an amber or dark spot at the nodus, of variable size.

PLATE 36   Pale-eyed Skimmers

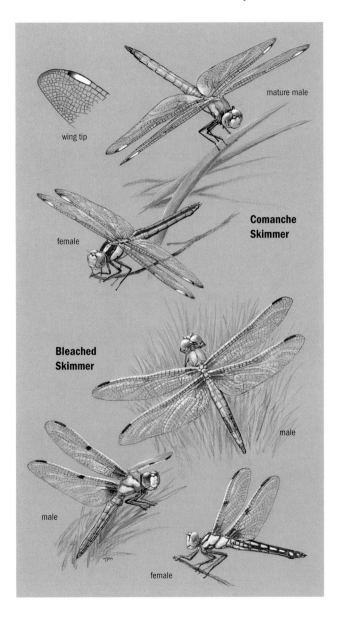

wing tip

mature male

female

Comanche Skimmer

Bleached Skimmer

male

male

female

PLATE 37   **Big Red Skimmers**

Flame Skimmer
(Libellula saturata)   PAGE 167

Length: 5 to 6 cm (2 to 2.5 in.)
Wing span: 8.5 to 9.5 cm (3.5 to 4 in.)

The male is a large, flame red dragonfly with rusty wash on the wings from the base to the nodus and brown streaks in the wing base. The female is tawny brown with rusty wash along the costal stripe out to the pterostigma and brown streaks in the wing bases.

Neon Skimmer
(Libellula croceipennis)   PAGE 166

Length: 5.5 cm (2 in.) Wing span: 8 to 9.5 cm (3 to 4 in.)

This dragonfly is similar to the Flame Skimmer (*L. saturata*), but the male's abdomen is bright neon red in color, the wings without brown streaks in the bases and with rusty wash only covering about a quarter of the wing, not reaching midwing except along the costal stripe. The female is like the female Flame Skimmer, but with no brown streaks and little wash of color at the wing bases.

PLATE 37   Big Red Skimmers

Flame Skimmer

male

female

male

Neon Skimmer

female

PLATE 38   Rock Skimmer and Clubskimmer

Red Rock Skimmer
(Paltothemis lineatipes)

PAGE 170

Length: 4.5 to 5.5 cm (2 in.)
Wing span: 9 to 9.5 cm (3.5 in.)

The mature male is mottled rusty red and black above, black below, with a rusty wash of variable extent on the basal half of the wings. The nondescript female has clear wings and is mottled with black, brown, and gray.

Pale-faced Clubskimmer
(Brechmorhoga mendax)

PAGE 171

Length: 5.25 to 6.25 cm (2 to 2.5 in.)
Wing span: 7.5 to 9 cm (3 to 3.5 in.)

The most noticeable feature is the white patch on the expanded, clublike tip of its long, narrow abdomen. The male looks gray on the wing except for the white patch, but has pale stripes on the thorax and the base of the abdomen and a pale face. The female looks like the male, but has a light brown wash on the wing tips.

PLATE 38 Rock Skimmer and Clubskimmer

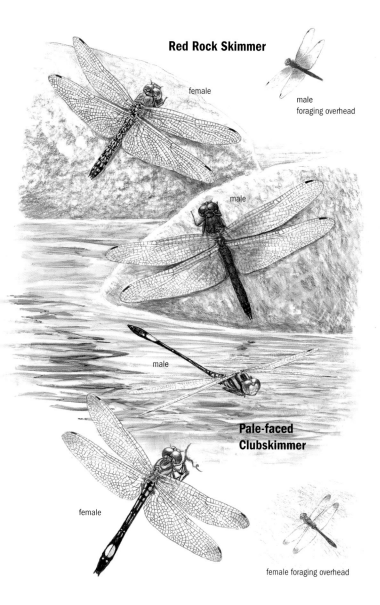

Red Rock Skimmer

female

male
foraging overhead

male

male

Pale-faced Clubskimmer

female

female foraging overhead

PLATE 39 Saddlebags and Marl Pennant

Black Saddlebags
(Tramea lacerata)   PAGE 174

Length: 5 to 5.5 cm (2 in.)
Wing span: 9.5 to 10 cm (3.5 to 4 in.)

The male is nearly all black with black patches resembling ink blots in the bases of the broad hind wings. The female is similar, usually with some yellow patches atop the abdomen.

Red Saddlebags *(Tramea onusta)*   PAGE 175

Length: 4 to 5 cm (1.5 to 2 in.)
Wing span: 8 to 9 cm (3 to 3.5 in.)

The dark brown wing patches have red veins. The male has an olive brown thorax and a red abdomen with black spots near the tip above. The female's body is dull, tawny and brown, but she also has red wing veins and wing patch.

Marl Pennant
(Macrodiplax balteata)   PAGE 172

Length: 4 cm (1.5 in.) Wing span: 7 cm (3 in.)

The male is nearly all black with oval, black patches in the bases of the hind wings and a white spot on the side of the head behind each eye. The female is mostly pale gray and dull yellow with similar black, oval patches in the wing.

PLATE 39   Saddlebags and Marl Pennant

ovipositing flight

Black Saddlebags

male

flight profile

female

Red Saddlebags

male

flight profile

female

Marl Pennant

male

female

male patrolling over water

PLATE 40 Gliders

These wind-riders have long, broad wings and short, tapered abdomens.

Wandering Glider
(Pantala flavescens)   PAGE 178

Length: 4.5 to 5 cm (2 in.)
Wing span: 8 to 9 cm (3 to 3.5 in.)

The abdomen is marked with orange and brown, brightest orange on the mature male. Wings may show a pale yellow wash at the base and tip, but are otherwise clear. The eyes of mature individuals are chestnut red.

Spot-winged Glider
(Pantala hymenaea)   PAGE 176

Length: 4.5 to 5 cm (2 in.)
Wing span: 8.5 to 9.5 cm (3.5 in.)

Its most distinctive feature is a small, round, dark spot low on the inner margin of each hind wing. Otherwise, it resembles the Wandering Glider *(P. flavescens)* in overall pattern and color, but is more tan and less orange.

PLATE 40   Gliders

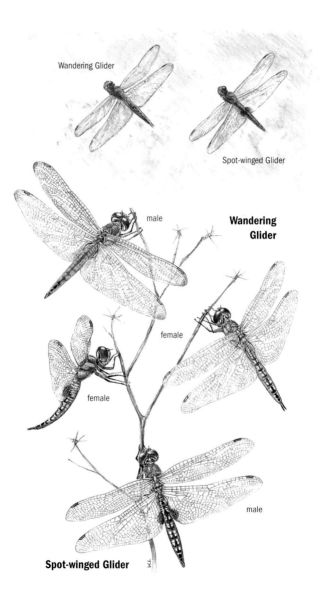

Wandering Glider

Spot-winged Glider

Wandering Glider

male

female

female

male

Spot-winged Glider

SKIMMERS, EMERALDS AND BASKETTAILS, AND CRUISERS (Libellulidae)

This is an immense, diverse group of species, the largest family of dragonflies in the world. Nearly two-thirds of the anisopterans in California are in this family. The most obvious, commonly shared feature of adults is that the large compound eyes are in fairly broad contact atop the head. This characteristic is shared by the darners *(Aeshnidae),* however. A more consistent, if less obvious, distinction of this family is that the triangles in the fore and hind wings are of very different shape and relative position. A distinctive feature of the larvae is the cupped labial mask covering the lower front of the face, shared only with the spiketail family (Cordulegastridae).

Cruisers *(Macromiinae)*

Cruisers are named for their fast and powerful flight and the males' habit of patrolling long beats down steams and rivers. The group is given family rank (Macromiidae) by many authors but here is classed as a subfamily. River cruisers *(Macromia)* are found worldwide, with about seven species in North America, one of these in California. Both sexes forage on the wing with great speed and agility, high and low, near and far from water. They are seldom seen perched and are difficult to follow—and even harder to catch.

WESTERN RIVER CRUISER **Macromia magnifica**
Pl. 24, Fig. 8
LENGTH: 7 to 7.5 cm (3 in.); **WING SPAN:** 9 to 10 cm (3.5 to 4 in.)
DESCRIPTION: This is a large, long-legged dragonfly with a single, complete, pale yellow vertical band on either side of the brown thorax, and two short, yellow stripes low on the front. The gray green to gray brown eyes come in contact atop the head. The face is pale yellow with two black bands, the topmost on the frons forming part of a T-spot atop the head. The blackish abdomen has pale yellow saddles on most segments and is club tipped (seg-

ments 7 through 10) on the male, cylindrical on the female. The squat, long-legged larva is a bottom sprawler with a distinct "horn" projecting rhinoceros-like from between the eyes.

SIMILAR SPECIES: It may be mistaken for dull darners *(Aeshna)* on the wing, and its color pattern is superficially like that of the Pacific Spiketail *(Cordulegaster dorsalis)* and the Black Petaltail *(Tanypteryx hageni)*, but none of these other species have just a single, pale yellow stripe on the side of the thorax.

BEHAVIOR: Males patrol low and fast along watercourses in the middle of sunny days with little to no wind. They are looking for females, which come to oviposit and do so by tapping their abdomens in the water of quiet pools with vegetated, overhanging banks. Away from water, both sexes forage over clearings at a wide range of heights, being almost constantly on the wing. They fly so fast that field marks are often difficult to see, and they can be mistaken for female darners. The male's clubbed abdomen is often carried slightly bowed in flight. They hang, when perched, like darners and spiketails.

DISTRIBUTION: The Western River Cruiser is widely distributed along streams and rivers from Kern and Inyo Counties northward to the Oregon border, primarily west of the Pacific Crest. A denizen of valleys and foothills, it is found from sea level to around 1,200 m (4,000 ft).

HABITAT: Large streams and small rivers are the preferred breeding habitats of this dragonfly. Occupied sites are typically stretches with a mix of shallow riffles and deeper pools. Foraging is done over clearings in wooded areas, occasionally far from water.

FLIGHT SEASON: This species is on the wing from mid-April to early September.

Emeralds and Baskettails *(Corduliinae)*

The generally somber-toned dragonflies in this subfamily primarily inhabit boreal and eastern North America, and only five of the 49 species found in the United States and Canada reach California. They are somewhat locally distributed within the state and can be hard to find, but they are occasionally common. Our species breed in ponds, lakes, or boggy streams. The anal loop in the hind wing is characteristically club shaped, without the pronounced "toe" of the foot-shaped anal loop seen on skimmers (Libellulinae) (fig. 11), and the emeralds and baskettails are often ranked together in a separate family (Corduliidae).

Common Emeralds and Striped Emeralds
(Cordulia and *Somatochlora)*

The emeralds are primarily boreal species, dark-bodied hawkers with vivid, emerald green eyes. The common emeralds *(Cordulia)* consist of but two species, one in North America, the other in Eurasia. Striped emeralds *(Somatochlora),* however, comprise some 40 species, 26 of them found in North America. Only three species of emeralds range south into the mountains of northern California.

AMERICAN EMERALD   *Cordulia shurtleffii*
Pl. 25

LENGTH: 4.5 to 5 cm (2 in.); **WING SPAN:** 6.5 cm (2.5 in.)

DESCRIPTION: A dark dragonfly, it looks mostly black in the field. The glowing, emerald eyes of the adult are the most distinctive feature (this color fades almost immediately after death). The teneral has brown eyes. The top of the face is metallic green. The thorax is metallic green and bronze, unpatterned, and quite hairy. The abdomen, spindle shaped in the male and cylindrical in the female, is black with inconspicuous blue green iridescence and a thin, whitish ring at the base of segment 3. The wings are clear but washed with amber brown on the teneral. In hand, the terminal appendages are distinctive. On the male the inferior appendage is most striking, bearing two lateral forks that in turn fork at their tips. The cerci of the female are less than 2 mm long. The larva (and exuvia) has a distinctive, dark stripe on the side of the thorax.

SIMILAR SPECIES: The Mountain and Ringed Emeralds *(Somatochlora semicircularis* and *S. albicincta)* are also dark bodied with bright green eyes, but they have yellow spots or stripes on the sides of the thorax and, in the latter species, narrow, white rings between the abdominal segments. The inferior appendage on male striped emeralds *(Somatochlora)* tapers to a dull point. The female Mountain Emerald has longer cerci (more than 3 mm) than those of the female American Emerald.

BEHAVIOR: Away from water, these emeralds typically hawk over meadows, bogs, and forest clearings, from near the ground to well overhead. Adults at breeding sites course low and fast with a darting flight over the water's surface. At rest, they hang vertically from perches on the outer branches of trees and shrubs. Tenerals disperse from emergence sites (logs and twiggy vegetation along lakeshores) to brushy clearings in nearby forest. There they may be found in good numbers perched low in the brush, taking short flights when disturbed.

DISTRIBUTION: This is the most widely distributed and frequently encountered emerald in the state, found around mountain lakes and bogs from the Oregon border southward to Fresno County in the Sierra Nevada and Mendocino County in the Coast Ranges. Elevations occupied in the Sierra Nevada range from about 1,200 to 2,700 m (4,000 to 9,000 ft). In northwestern California, where conifer forests reach the coast, they are found at lower elevations, occasionally near sea level.

HABITAT: The American Emerald breeds in lakes but often forages over other habitats such as boggy meadows and forest clearings.

FLIGHT SEASON: Known dates in California range from late May to early September.

MOUNTAIN EMERALD   *Somatochlora semicircularis*
Pl. 25

LENGTH: 5 to 5.5 cm (2 in.); **WING SPAN:** 6.5 to 7 cm (2.5 to 3 in.)

DESCRIPTION: This dragonfly is similar in overall pattern and color to the American Emerald *(Cordulia shurtleffii),* but with two yellow spots (a larger, oval anterior one and a smaller, round posterior one) on the side of the thorax. The cerci of the male curve outward at midlength and then back together at their tips to form "semicircles." The female has yellow lateral spots atop abdominal segment 3 and cerci that are 3 mm or more in length.

SIMILAR SPECIES: Other emeralds, especially the American Emerald, look similar on the wing. In the hand or viewed at close range while perched, the American Emerald lacks yellow spots on the thorax, and the male's cerci are not pincerlike. The female American lacks yellow spots on abdominal segment 3 and has shorter cerci; the Ringed Emerald *(S. albicincta)* has distinctive, white rings on its abdomen.

BEHAVIOR: Males patrol and forage low over dense emergent vegetation of bogs, meadows, and lakeshores. The long abdomen is sometimes curled downward in flight.

DISTRIBUTION: This sleek emerald can be found in modest numbers at mountain meadows and swamps in the North Coast Ranges southward to Mendocino County and in the Cascade-Sierran

Province southward to Tulare County. It has been recorded at elevations ranging from 1,000 m (3,200 ft) in the northwestern corner of the state to 3,300 m (11,000 ft) in the mountains.

HABITAT: The Mountain Emerald is typically found over boggy meadows or low, dense emergent vegetation (grasses, sedges) bordering ponds, lakes, and streams. Nonbreeding individuals also may be found around forest clearings and dry meadows.

FLIGHT SEASON: The brief season in California is from mid-June through August.

RINGED EMERALD *Somatochlora albicincta*
Pl. 25

LENGTH: 4.5 to 5 cm (2 in.); **WING SPAN:** 6.5 cm (2.5 in.)

DESCRIPTION: Like our other emeralds, this is a dark dragonfly with green eyes on the mature adult, brown on the immature. The metallic green thorax has a single, pale mark on the side. Most distinctive are the narrow, incomplete white rings between abdominal segments.

SIMILAR SPECIES: This species is likely to be confused with American and Mountain emeralds *(Cordulia shurtleffii* and *S. semicircularis)* in California, but both of these lack the white abdominal rings.

BEHAVIOR: Males patrol low along the water's edge.

DISTRIBUTION: Only a few specimens of this distinctive species have been collected in California, at or near high mountain lakes in Shasta, Siskiyou, and Plumas Counties. Recorded elevations range from 1,700 to 2,000 m (5,500 to 6,700 ft).

HABITAT: This emerald typically occupies lakes, ponds, or streams with sparse emergent vegetation and forested margins.

FLIGHT SEASON: In California it has been found emerging in mid-June, with adults collected in July and August.

Baskettails *(Tetragoneuria)*

Baskettails are dragonflies of wooded ponds, lakes, and quiet riparian backwaters and most numerous in the humid climes of eastern North America. In the western United States they are more or less restricted to the wetter, temperate wooded areas of Idaho, Washington, Oregon, and northern California. Species in this genus are typically clad in somber earth tones of black, olive, brown, and butterscotch and are difficult to distinguish in the field. Two species occur in northern California, only one of which is widely distributed. The baskettails are often included in the Eurasian genus *Epitheca.*

BEAVERPOND BASKETTAIL *Tetragoneuria canis*
Pl. 26

LENGTH: 4.5 to 5 cm (2 in.); **WING SPAN:** 6.5 cm (2.5 in.)

DESCRIPTION: These are dark, medium-sized dragonflies. The abdomen is flattened and so appears narrow in side view. Viewed from above, the abdomen is spindle shaped (widest in the middle, tapering toward the base and tip). The body is mostly brown and black, with a row of yellow orange spots on each side of the abdomen, and the face is yellow. The thorax is heavily coated with hairs. Eye color varies from gray brown in the teneral to blue gray, teal blue, or bright jade green in the mature adult (brighter in the male). The dorsal surface of the frons is mostly pale. The hind wings have small and inconspicuous dark basal patches. The epithet *canis,* Latin for "dog," refers to the cerci of the male, which somewhat resemble a dog's head in profile (the ventrally oriented tip of each appendage being the "muzzle" and the small dorsal projection at its angle being the "ear"). The cerci of the female are less than 3 mm in length, usually about 2 mm.

SIMILAR SPECIES: The Spiny Baskettail *(T. spinigera)* is virtually identical except for the structure of its terminal appendages and its darker frons. Flying individuals are impossible to distinguish. Extremely close views of a perched baskettail with close-focusing binoculars might allow tentative identification, but in nearly all cases species identification requires in-hand examination. The superior appendage of the male Spiny Baskettail is somewhat upturned at the tip but lacks a dorsal projection; it does have a sharp ventral spine near its base (rarely reduced or absent). The terminal appendages of female Beaverpond Baskettail are shorter than those of female Spiny Baskettail which are more than 3 mm.

BEHAVIOR: Beaverpond Baskettails are easily observed hawking along trails and through clearings within about 1 m (3 to 4 ft) above the ground. They often seem relatively unconcerned by passersby and will glide right by a still observer standing in their path. When resting, they hang low in weedy vegetation. Males patrol the edges of willow-lined ponds and quiet riparian backwaters. The females carry egg masses under the tip of the abdomen in basketlike structures (hence the common name), then lay their eggs in gelatinous strings on vegetation at the water's surface.

DISTRIBUTION: This species is of local occurrence in the valleys and foothills of northern California southward to Marin, Solano, Sacramento, and Calaveras Counties. It is known from near sea level in the San Francisco Bay Area to 1,400 m (4,500 ft) in Modoc County. It is often quite common where it is found.

HABITAT: Found along riparian corridors and at ponds, most often in the valleys of larger streams and rivers, it forages in openings or along trails and roads in wooded areas.

FLIGHT SEASON: This species emerges early—there are even a few March records for the state—typically in April and May, with few records after June. The latest known flight date for California is July 21.

SPINY BASKETTAIL   *Tetragoneuria spinigera*
Pl. 26, Fig. 6

LENGTH: 4.5 cm (2 in.); **WING SPAN:** 6 to 6.5 cm (2.5 in.)

DESCRIPTION: This dragonfly is similar to the Beaverpond Baskettail *(T. canis)* (see the "Description" section for that species) but has a darker frons (T spot). The larva is quite distinctive among California dragonflies in having very long lateral spines on abdominal segment 9 that extend back beyond the tip of the abdomen (fig. 6).

SIMILAR SPECIES: See the "Similar Species" section for the Beaverpond Baskettail. The Four-spotted Skimmer *(Libellula quadrimaculata)*, a common dragonfly of mountain lakes in California, bears a superficial resemblance to the baskettails, but it has a paler face, dark nodal wing spots, a stocky abdomen that gradually tapers at the tip and is much different in behavior (it frequently perches on low vegetation near water, from which it sallies out for food).

BEHAVIOR: Adults typically forage some distance from water in openings and along roads in forested areas. When they visit breeding lakes, they are most frequent out over open water far from shore, and breeding activity is reported to occur in late afternoon and at dusk. Larvae take 2 years to mature. Emergence is explosive, with thousands of individuals packed close together on shoreline vegetation emerging virtually simultaneously. Upon taking wing, the tenerals fly away from the lake to forage in clearings in the forest. Their foraging habits are similar to those of Beaverpond Baskettails.

DISTRIBUTION: The Spiny Baskettail is known from only two locations in the state: Donner Lake, Nevada County, and Blue Lake, Lassen County, both at an elevation of approximately 1,800 m (6,000 ft). This species is almost certainly more widely distributed in the state than the few records suggest, but its furtive behavior hinders detection.

HABITAT: This baskettail has been found at mountain lakes with wooded shores and sparse emergent vegetation along the shoreline. Exuviae have been found in masses in low, shoreline sedges and on the lower sides of half-submerged logs.

FLIGHT SEASON: The few records for California are in June and July. Mass emergence apparently occurs in late May or early June. Mid-June is perhaps the best time to look for mature adults.

Skimmers *(Libellulinae)*

The skimmers include some of our flashiest and most familiar dragonflies. More than 100 species in this subfamily occur in North America, 38 of these in California. Many are pond species, but some occur along streams and rivers. Mature males and females are often quite different in appearance, the male typically more brightly colored on the abdomen in red, white, or blue, the white and blue colors coming from pruinescence that develops with age. A number of species have conspicuous black or brown patches in the wings. The most distinctive morphological characteristic shared by all our species is a foot-shaped anal loop in the hind wing (fig. 11). The group is often given family status (Libellulidae).

Whitefaces *(Leucorrhinia)*

Whitefaces are a distinctive genus of 17 species primarily inhabiting the boreal and mountain regions of the Northern Hemisphere. Four of the seven North American species occur in California at bogs and marshy ponds and lakes in the northern third of the state and southward in the Sierra Nevada to Tulare County. Adults fly low near the water, perching on the ground or low plants, including floating vegetation. They are small and dark bodied, with a striking, white face. The thorax and parts of the abdomen are marked with red or yellow, these colors fading somewhat with age. The legs are black. The pterostigma is dark, short, and thick. The wings are clear except for some small, dark areas at their roots and occasional basal tints of brown or amber.

HUDSONIAN WHITEFACE   *Leucorrhinia hudsonica*
Pl. 27, Fig. 13
LENGTH: 3 to 3.5 cm (1 to 1.5 in.); **WING SPAN:** 5 to 6 cm (2 to 2.5 in.)
DESCRIPTION: The color patterns of the male and female are similar: white face; black body with red (or yellow on some females and the immature male) stripes and irregularly shaped blotches on the thorax and basal abdominal segments (1 through 3); and a red or yellow dorsal spot on each of segments 4 through 7. In hand, the male epiproct appears parallel sided and slightly forked and is nearly three-fourths the length of the downturned cerci. The lobes of the vulvar lamina of the female are fairly large. One row of cells is subtended by the radial planate.
SIMILAR SPECIES: The mature male is usually distinguished from our other male whitefaces by the yellow or red spots on segments 4 through 7. These spots may be so small as to be nearly impossible to see except at close range, so caution is advised. The females of all other whiteface species are similar and hard to tell apart except in hand. The lobes of the vulvar laminae of our other species are all quite small relative to those of the Hudsonian Whiteface female.
BEHAVIOR: These are rather unwary whitefaces that fly in sunny, open habitats. Males usually perch low on the ground or on the short stems of sedges and other low plants. The female, guarded by the male hovering nearby, oviposits in shallow water with emergent vegetation by repeatedly tapping the surface with the tip of her abdomen.
DISTRIBUTION: This is a boreal species found across the taiga of Canada and Alaska, across the northern border states, and southward in the high mountains of the West, including the North Coast Ranges to Mendocino County and the Cascade-Sierran Province southward to Tulare County. It has been recorded at elevations ranging around 900 to 2,700 m (3,000 to 9,000 ft) within California.
HABITAT: Boggy meadows and the boggy margins of ponds and lakes are frequented by this species.

FLIGHT SEASON: Its flight season is fairly early for a montane dragonfly, from late May through August.

CRIMSON-RINGED WHITEFACE   *Leucorrhinia glacialis*
Pl. 27, Fig. 13

LENGTH: 3.5 to 4 cm (1.5 in.); **WING SPAN:** 5.5 to 6 cm (2 to 2.5 in.)

DESCRIPTION: The male has the typical whiteface pattern of a white face and black and red markings on the thorax and basal abdominal segments, but it usually has segments 4 through 10 all black. The young male has the pale markings yellow instead of red. In hand, note that the epiproct is only about half as long as the upturned cerci, and the anterior arm of each hamule has a small hook at the tip. The female, like our other whiteface females, has red or yellow spots atop abdominal segments 4 through 7. The vulvar lamina has very short lobes separated by a slight indentation. Both sexes show two rows of cells subtended by the radial planate.

SIMILAR SPECIES: The Red-waisted Whiteface *(L. proxima)* is identical to the Crimson-ringed Whiteface in the field and requires in-hand differentiation. The Red-waisted Whiteface has only one row of cells subtended by the radial planate, the male's epiproct is about two-thirds as long as the cerci, and the female has a slightly longer and more distinctly lobed vulvar lamina. Some male Hudsonian Whitefaces *(L. hudsonica)* have reduced pale spots atop abdominal segments 4 through 7 and can be mistaken for this species unless closely inspected. The female Hudsonian Whiteface, distinguishable only in hand, has a single row of cells subtended by the radial planate, and long, distinct lobes on the vulvar lamina.

BEHAVIOR: They are often found perched on bushes in the dappled sunlight of forest clearings adjacent to breeding ponds and bogs. They are not averse to perching in shady areas and are somewhat difficult to follow as they move from sun to shadow. Males hang out on shoreline brush and weeds waiting to capture females out over emergent vegetation and carry them back to shore for breeding. Oviposition is like that of the Hudsonian Whiteface.

DISTRIBUTION: This species occurs at relatively high elevations, even for a whiteface, between 1,700 and 2,700 m (5,500 and 9,000 ft) in the Cascade-Sierran Province, including the Warner Mountains, from the Oregon border southward to Mariposa County. It is widely distributed across the northern states and Canada.

HABITAT: This species seems partial to bogs, ponds, and lakes with emergent vegetation in the shallow margins and brush and open forest along the shore.

FLIGHT SEASON: Its typical season is relatively short, from late June to early September, probably because of the higher elevations it occupies. There is one report of emergence in late May.

RED-WAISTED WHITEFACE   *Leucorrhinia proxima*
Pl. 27, Fig. 13
LENGTH: 3.5 cm (1.5 in.); **WING SPAN:** 5 to 5.5 cm (2 in.)

DESCRIPTION: The mature male is patterned like the Crimson-winged Whiteface *(L. glacialis):* black bodied with a red patchwork pattern on the thorax and basal abdominal segments, the rest of the abdomen solid black; it may have tiny, red midlines on top of some abdominal segments, which are scarcely if at all visible except in hand. In hand, note the epiproct that is two-thirds the length of the cerci, the one row of cells subtended by the radial planate, and the shape of the hamules (fig. 13). The young male is yellow where the mature male is red. The female is patterned in red or yellow and black, like other female whitefaces, and is difficult to distinguish. The vulvar lamina has short, rounded lobes in contact at their bases.

SIMILAR SPECIES: Without in-hand examination, the female is not safely distinguished from the Hudsonian Whiteface *(L. hudsonica)* female, and neither sex is distinguishable from the Crimson-ringed Whiteface *(L. glacialis).* The female Hudsonian Whiteface has a prominently lobed vulvar lamina. The Crimson-ringed typically has two rows of cells subtended by the radial planate, the male epiproct about half the length of the cerci, the male hamules different, and the vulvar lamina of the female only shallowly lobed (fig. 13).

BEHAVIOR: Red-waisted Whitefaces behave like other whitefaces. They frequent perches atop bushes and fern fronds on the margins of breeding sites and in forest clearings nearby. Females oviposit at pools on the bog.

DISTRIBUTION: This is another boreal species, recently discovered at Willow Lake, Plumas County, elevation 1,700 (5,500 ft), its only known location in the state. The nearest known populations are in Washington State, but it should be looked for elsewhere in California and Oregon where appropriate habitat exists.

HABITAT: The breeding habitat at Willow Lake is a sphagnum bog, including floating islands, bordering a mountain lake. Ponds and lakes with marshy or boggy borders are occupied elsewhere.

FLIGHT SEASON: This species has been seen from mid-June through mid-August at Willow Lake.

DOT-TAILED WHITEFACE *Leucorrhinia intacta*

Pl. 27, Fig. 13

LENGTH: 3.5 cm (1.5 in.); **WING SPAN:** 5.75 cm (2 in.)

DESCRIPTION: The young male and many females have the usual whiteface pattern: white face, dark eyes, and black body with a patchwork of yellow spots and stripes on the thorax and anterior abdominal segments, segments 4 through 7 each with a yellow dorsal spot. This pattern fades fairly quickly with age, especially on the male, leaving the thorax and abdomen mostly black with a bright yellow, square patch atop abdominal segment 7. The best in-hand characteristic for the male Dot-tailed is the epiproct, which has broadly diverging sides so that it appears much wider at the tip than at the base, the tip noticeably forked. The vulvar lamina of the female has two small, widely spaced lobes.

SIMILAR SPECIES: The adult Hudsonian Whiteface *(L. hudsonica)* retains dorsal spots of color (yellow on the female, red on the male) on abdominal segments 4 through 6, as well as the yellow or red markings on the thorax. The bright female and the very young male Dot-tailed Whiteface are quite similar to other whiteface females, requiring in-hand examination in most cases. On other whitefaces, males have the epiproct more or less parallel sided, with a shallow fork at best; females have the lobes of the vulvar lamina in contact, or nearly so, at their bases. The male Black Meadowhawk *(Sympetrum danae)* is also mostly black with a (usually less conspicuous) yellow spot on segment 8, but it has a dark face and a clear hind wing base.

BEHAVIOR: Male Dot-tailed Whitefaces seem to like open, sunny perch sites, either on the ground or out in the water some distance from shore on emergent or floating vegetation. They maintain small territories at breeding sites, awaiting the infrequent visits of females. Males guard ovipositing females by hovering nearby and harassing intruders (i.e., other males).

whitefaces

| Dot-tailed | Hudsonian | Crimson-ringed | Red-waisted |

hamules (side views)

| Dot-tailed | Hudsonian | Crimson-ringed | Red-waisted |

vulvar laminae (bottom views)

---

meadowhawks

| Variegated | Cardinal | Red-veined |

| Cherry-faced | White-faced | Striped |

vulvar laminae (bottom views)

| Western | Saffron-winged | Black | Yellow-legged |

vulvar laminae (shaded, side views)

Figure 13. Structural characteristics for identifying whitefaces and meadowhawks.

DISTRIBUTION: This is a wide-ranging whiteface in California, found at elevations from about 1,100 to 2,600 m (3,500 to 8,500 ft) in the Cascade-Sierran Province southward to Fresno County, in the Great Basin Province in Lassen and Modoc Counties, and in the North Coast Ranges southward to Glenn County.

HABITAT: This species occurs at boggy and marsh-bordered ponds and lakes, including farm and stock ponds, and even along slow-moving streams. It can be found foraging away from breeding sites in wet meadows.

FLIGHT SEASON: It flies from late May into September, a bit longer than our other whitefaces.

Meadowhawks (Sympetrum)

This is a large genus of small to midsized dragonflies of wide occurrence, primarily in the Northern Hemisphere. In North America, meadowhawks are most abundant and diverse in the northern interior of the continent. They frequently outnumber all other dragonflies in open country, where they typically forage by sallying out from exposed perches. Ten of the 15 North American species are found in California. Meadowhawk diversity is much greater in the northeastern part of the state than in the south. Six or seven species may be found at a single mountain meadow in the southern Cascade Range, whereas only three species have been recorded in vast San Bernardino County. Although a few species fly in spring, most are late-summer and fall breeders. Reddish coloration is typical of the mature individuals, at least the males, of all our species but one. Some of our larger, more widely distributed species—Variegated, Cardinal, and Red-veined Meadowhawks (*S. corruptum, S. illotum,* and *S. madidum*)—share certain characteristics and were at one time considered members of a separate genus (*Tarnetrum*).

BLACK MEADOWHAWK   *Sympetrum danae*
Pl. 28, Fig. 13
LENGTH: 2.75 to 3 cm (1 in.); **WING SPAN:** 4.5 to 5 cm (2 in.)
DESCRIPTION: A small, black or black-and-yellow dragonfly. The young male resembles the female, which is black with yellow spots on the sides of the thorax and has the abdomen yellowish above and black on the sides and undersurface. The black areas of the abdomen encroach upon the dorsal yellow areas with age, es-

pecially on the mature male, which is virtually all black, including the face; the face is yellowish on the female. A small pair of yellow spots may linger on the upper surface of abdominal segment 8 on the male. The pterostigma is small and black above. The wings of the female have a variable, amber wash at the bases and along the leading edges, especially at the nodus. The vulvar lamina is a short spout.

SIMILAR SPECIES: The black face of the male easily distinguishes it from the other small, blackish dragonflies that share its habitat: the whitefaces (*Leucorrhinia*). Our only meadowhawk with a somewhat similar pattern on the young male and the female is the Western Meadowhawk (*S. occidentale*), which is typically larger and has more yellow on the thorax (mostly yellow in front and with a boot-shaped yellow mark on the side where the Black Meadowhawk has spots). The vulvar lamina of the female Western Meadowhawk is not prominently spoutlike.

BEHAVIOR: Black Meadowhawks forage low in open areas—bogs, meadows, marshes, and lakeshores. Oviposition, done in tandem or by the female alone, is accomplished in a number of ways, either by tapping the water with the tip of the abdomen, dropping eggs from the air, or inserting the spoutlike vulvar lamina into wet mud.

DISTRIBUTION: This circumboreal species is known in California only from elevations over 1,500 m (5,000 ft) in the Cascade-Sierran and Great Basin Provinces southward to Mono County and the North Coast Ranges in Trinity County.

HABITAT: It is found at bogs, wet meadows, and the marshy borders of mountain lakes.

FLIGHT SEASON: It has a short season, mid-June through September, at the high elevations it occupies.

WESTERN MEADOWHAWK　　*Sympetrum occidentale*

Pl. 28, Figs. 3, 13

LENGTH: 3 to 4 cm (1 to 1.5 in.); **WING SPAN:** 4.5 to 6 cm (2 to 2.5 in.)

DESCRIPTION: This is a small to medium-sized meadowhawk usually with a wash of amber or rusty brown on the basal half of the wings (faint on the fore wing), darkest near the midwing and fading toward the base. The wash of color may be quite pale or even absent on the female, however. The young male and female have the body mostly yellow (tan or greenish yellow on the face, golden yellow on the front of the thorax and the top of the abdomen), with the top of the eyes brown and the following black marks: marbled lines extending upward from the leg bases, outlining a yellow, boot-shaped mark on the side of the thorax; a thick, lateral stripe on the abdomen; and spots atop abdominal segments 8 and 9. The yellow part of the abdomen turns red and the thorax becomes more or less brown on mature individuals. The male's cerci are red (yellow when immature), the epiproct black. The pterostigma is brown (reddish on the mature male), and the legs are black.

SIMILAR SPECIES: The Cherry-faced Meadowhawk *(S. internum)* can have a similar amber or rust wash on the wings, but this is typically darker near the base and fading outward, rather than darker toward the midwing as in the typical Western. The Cherry-faced is best distinguished by its plain thorax without black markings. A female Western lacking a wash of color in the wings is also best distinguished from other female meadowhawks (e.g., Black *[S. danae]*, Cherry-faced) by the pattern of black and yellow on the side of the thorax. Other species with a red or brown wash on the basal half of the hind wing—Widow, Flame, Neon, and Red Rock Skimmers *(Libellula luctuosa, L. saturata, L. croceipennis,* and *Paltothemis lineatipes)*—are much bigger.

BEHAVIOR: Areas of dense, low emergent vegetation seem preferred for ovipositing, which is done in tandem. Nonbreeding foragers may form large congregations near or far from water, perched on marsh plants or in rows along fence lines and telephone wires.

DISTRIBUTION: Sometimes considered conspecific with the Band-winged Meadowhawk *(S. semicinctum)* of eastern North America, this species is widespread and common in many parts of the interior West. In California, it is found primarily in the northeast quarter, southward in the Sierra Nevada to Kern County and in the Great Basin to Inyo County. It has also been found in modest numbers in the Central Valley and along the coast southward to Merced and Santa Cruz Counties. The elevations occupied range from sea level to over 2,100 m (7,000 ft).

HABITAT: The Western Meadowhawk breeds at boggy meadows, marshes, ponds, lakes, slow streams, and sloughs bordered by grasses and sedges. Nonbreeding individuals may be common along roadsides bordering croplands, pastures, or marshes.

FLIGHT SEASON: Dates range from April to September, but most are in June through August.

SAFFRON-WINGED MEADOWHAWK
Sympetrum costiferum
Pl. 29, Fig. 13

LENGTH: 3 to 4 cm (1 to 1.5 in.); **WING SPAN:** 5 to 6 cm (2 to 2.5 in.)

DESCRIPTION: The most distinctive features are along the leading edges of the wings: The veins in the costal stripe are amber aging to red, with a saffron yellow wash on the membrane between these (fading in the mature male). The long pterostigma is yellow, aging to orange or red and bordered fore and aft with black lines (veins). The face is pale yellow to yellow olive, aging red in the male. The immature male and female have the thorax yellow olive in front and yellow on the sides, with some variable (usually minimal to absent) black marbling. The thorax of the mature male is mostly brown. The abdomen is dark below and yellow aging to brick red above, especially in the male, with some black low on the side and along the dorsal midline of segments 7 through 9 (these later marks are of variable extent, from scant to moderate). The male's cerci and epiproct are yellow, aging to red. The legs are black with a tan stripe on the rear margin.

SIMILAR SPECIES: The Variegated, Cardinal, and Red-veined Meadowhawks *(S. corruptum, S. illotum,* and *S. madidum)* often have yellow to red veins in the costal stripe and/or a wash of amber or rust in this region, but all are more robust, have white or yellow spots or stripes on the side of the thorax, and have the pterostigma centrally or entirely dark. Western Meadowhawks *(S. occidentale)* with little wash of amber in the wings have dark pterostigma and more extensive black marbling on the sides of the thorax and abdomen. The Yellow-legged Meadowhawk *(S. vicinum)* has pale legs.

BEHAVIOR: Males make brief patrols with frequent hovering over emergent vegetation along the shoreline, then perch in sunny spots on wrack-strewn ground and low sedges. Oviposition occurs in tandem over open water near shore. Nonbreeding individuals may forage some distance from water in open country.

DISTRIBUTION: This species ranges across Canada and the northern United States but is restricted in California to the Cascade-Sierran and Great Basin Provinces from Siskiyou County southward to Inyo County. It occurs at elevations between 900 and 2,300 m (3,000 and 7,500 ft).

HABITAT: The Saffron-winged Meadowhawk frequents springs, marshes, and the borders of lakes and ponds, often with patchy or sparse vegetation, and sand, gravel, or drifted debris along the shoreline. It is found at both alkaline lakes and acidic bogs and occasionally at creeks or sloughs.

FLIGHT SEASON: It flies in late summer and fall, from late June into October.

YELLOW-LEGGED MEADOWHAWK   *Sympetrum vicinum*
Pl. 28, Fig. 13

LENGTH: 3 to 3.5 cm (1 to 1.5 in.); **WING SPAN:** 4.5 cm (2 in.)

DESCRIPTION: A small, slender meadowhawk, this species is rather plain with few dark markings. The legs are pale and yellowish, aging to red brown in males. The immature male and female are yellow and tan or gray, with few if any black markings on the abdomen. The wings are clear with a small amount of amber (redder in the mature male) at the very base. The mature male has a brown thorax, red abdomen, and red on its face. The mature female reddens on the top of the abdomen, too. The female has an abdomen extremely swollen at the base (viewed from the side), and a greatly expanded, trumpet-shaped vulvar lamina.

SIMILAR SPECIES: The Saffron-winged Meadowhawk *(S. costiferum)* can be nearly as plain on the thorax and abdomen, but it has a paler pterostigma and black on its legs, and the female lacks an enlarged vulvar lamina. The Striped Meadowhawk *(S. pallipes)* may have pale legs but has a striped thorax. The female White-faced Meadowhawk *(S. obtrusum)* has black legs and black on the sides of the abdomen.

BEHAVIOR: They often forage away from water in wooded areas in the morning; in the afternoon, pairs come to oviposit in tandem in shallow water or adjacent shore. They usually fly low and can forage at fairly low temperatures.

DISTRIBUTION: Widespread and fairly common in eastern North America and the Pacific Northwest, this small dragonfly has been found only at a few locations in the Cascade-Sierran Province as far south as Placer County at an elevation range of about 1,100 to 1,700 m (3,500 to 5,500 ft). It no doubt occurs elsewhere in California but has been overlooked, perhaps because of its late flight season.

HABITAT: It is found in meadows and vegetation bordering ponds and lakes, often in wooded areas.

FLIGHT SEASON: This is a late-flying species, recorded in California in September and October (it flies into November elsewhere).

CHERRY-FACED MEADOWHAWK   *Sympetrum internum*
Pl. 29, Fig. 13

LENGTH: 3 to 3.5 cm (1 to 1.5 in.); **WING SPAN:** 5 to 6 cm (2 to 2.5 in.)

DESCRIPTION: Its tawny yellow face becomes cherry red in the mature male. The anterior wing veins are orange red, and the bases of the wings often have an amber wash of variable extent; the pterostigma is brown. The thorax is plain brown above and yellower on the sides, becoming all reddish brown on the mature male. The abdomen is tan above, becoming bright red on the mature male and bordered on the sides by a thick, wavy black line. The legs are black.

SIMILAR SPECIES: It is most similar to the White-faced Meadowhawk *(S. obtrusum)*, which has a pale face and black wing veins. Cherry-faced Meadowhawks with extensive amber wash at the base of the wings are somewhat similar to the Western Meadowhawk *(S. occidentale)*, but lack the black marbling on the thorax of the latter species. The Saffron-winged Meadowhawk *(S. costiferum)* has a pale pterostigma and partly pale legs. The Red-veined Meadowhawk *(S. madidum)* has spots or stripes on the thorax and lacks extensive black on its abdomen.

BEHAVIOR: Like other meadowhawks, they fly low and forage in weedy and emergent vegetation. Pairs, often congregating at preferred sites, oviposit in tandem, females dropping eggs from the air on seasonally dry wetland substrates such as the receding shoreline of ponds in late summer.

DISTRIBUTION: A widespread and common species across Canada and the northern states, the Cherry-faced Meadowhawk is little known in California, with records from scattered locations in the Sierra Nevada southward to Fresno County and the Great Basin Province from Modoc County to Mono County, at elevations from about 1,200 to 2,200 m (4,000 to 7,200 ft).

HABITAT: Freshwater marshes, marshy borders of streams, ponds, and roadside ditches, and boggy meadows are frequented by this species. It has been found foraging in roadside weeds bordering freshwater marshland in the company of other meadowhawk species.

FLIGHT SEASON: It flies in late summer and autumn, mid-June through September, when aquatic habitats are beginning to dry up.

WHITE-FACED MEADOWHAWK　　*Sympetrum obtrusum*

Pl. 29, Fig. 13

LENGTH: 3 to 4 cm (1 to 1.5 in.); **WING SPAN:** 4.5 to 5.5 cm (2 in.)

DESCRIPTION: This is a slender, delicate species. The face is pale yellow in the immature individual, becoming white on the adult. The wings are clear with black veins, and they have dark brown pterostigmata. Otherwise, the immature male and most females have rather plain, tan and golden yellow bodies, and abdomens with thick, wavy, black lateral stripes. The abdomen becomes bright red above on the mature male (some females, too), and the thorax is brown. The legs are mostly black.

SIMILAR SPECIES: The Cherry-faced Meadowhawk *(S. internum)* is very similar, but it has rusty anterior wing veins and a darker face (tan, olive, or red depending on age and sex). The Striped Meadowhawk *(S. pallipes)* has pale stripes on its thorax. The Yellow-legged Meadowhawk *(S. vicinum)* has pale legs and lacks the black lateral stripe on the abdomen.

BEHAVIOR: They fly low over sunlit grasses and sedges and are fairly easily approached. Females oviposit, in tandem or with the male hovering on guard nearby, by dropping eggs onto mud or vegetation in drying meadows, marshes, or the receding shoreline of ponds or bogs.

DISTRIBUTION: This a common, sometimes abundant, dragonfly at elevations ranging from 1,200 to 2,400 m (4,000 to 8,000 ft) in the Cascade-Sierran Province from Siskiyou County southward to Tuolumne County. Its North American range is similar to that of the Cherry-faced Meadowhawk.

HABITAT: These dainty dragonflies swarm in boggy meadows and the dense emergent vegetation around mountain lakes and ponds. Nonbreeding individuals forage to some extent in sunny, weedy clearings in adjacent forests.

FLIGHT SEASON: Although collected once in the state in late May, this is typically a late-summer and fall flyer, July through October.

STRIPED MEADOWHAWK　　*Sympetrum pallipes*

Pl. 31, Fig. 13

LENGTH: 3.5 to 4 cm (1.5 in.); **WING SPAN:** 6 to 6.5 cm (2.5 in.)

DESCRIPTION: This is a rather plain, brown dragonfly with two creamy white stripes on the side and two whitish dashes on top of the thorax, and when mature, the top of the abdomen is tomato red. The face is a pale cream or tan, and the top of the eyes is a rich brown at maturity. Each wing may be amber tinted or rusty at the very base and have a red costa, but it is typically clear, with the pterostigma reddish brown and slightly paler at the ends. The abdominal segments have a black lateral line, variable in extent and often faint, and become white pruinescent below (this is especially noticeable on the thicker basal segments). The leg color varies from tan to mostly black.

SIMILAR SPECIES: The Variegated Meadowhawk *(S. corruptum)* has similar striping on the thorax, but it also has a distinctive row of black-rimmed white spots on the side of the abdomen. The Red-veined Meadowhawk *(S. madidum)* also has pale stripes on the side (but not the top) of the thorax, but the leading edge (costal area) of the wings is washed with amber or rust, and the mature male has a reddish face and thorax. The White-faced Meadowhawk *(S. obtrusum)* lacks pale stripes on the thorax.

BEHAVIOR: Males at breeding sites perch close to the ground on grass and sedge stems or low brush. Pairs oviposit in tandem or females do so alone, dropping eggs on the vegetation of exposed shoreline or dried ponds to hatch when the water level rises. Nonbreeding individuals forage from near the ground to 6 m (20 ft) up along fence lines or in brush and trees in open country or woodland clearings.

DISTRIBUTION: A common species throughout much of the western United States and southwestern Canada, the Striped Meadowhawk is found throughout the northern two-thirds of California. It has also been collected as far south as the San Bernardino Mountains, San Bernardino County. It ranges in elevation from sea level to over 2,400 m (8,000 ft) in the Sierra Nevada.

HABITAT: This species breeds in a variety of habitats: freshwater marshes, wet meadows, bogs, ponds and lakes (including temporary ones) with extensive exposed margins at low water levels, and occasionally streams with similar habitat edges. Nonbreeding individuals are often found far from water along roadsides, in clearings in forested areas, in open grasslands, and even in suburban yards.

FLIGHT SEASON: It may be found on the wing from late April to November at lower elevations; its season is of shorter duration in the mountains.

RED-VEINED MEADOWHAWK   *Sympetrum madidum*
Pl. 30, Fig. 13

LENGTH: 4 to 4.5 cm (1.5 to 2 in.); **WING SPAN:** 6.5 to 7 cm (2.5 to 3 in.)

DESCRIPTION: This is a fairly robust meadowhawk with rust-colored anterior wing veins and an amber or rust wash on the leading half of the wing. The pterostigma is dark brown to reddish brown above. The face is yellow on the immature individual and the female, but red on the mature male. The thorax is a rich brown (to red on the male) with two whitish stripes on the side, fading to round, white spots at their lower ends on the mature male. The female and the immature male have the abdomen yellow tan with two thin, parallel, black lateral lines, the uppermost broken into a series of wavy dashes, and small, black marks atop segments 8 and 9. The abdomen is a rich, rufous red on the mature male (the intensity of the red color is apparently somewhat temperature dependent). Its relatively robust stature, pale dots on the side of the thorax, and two rows of wing cells subtended by the radial planate are features shared with other *"Tarnetrum"* species.

SIMILAR SPECIES: The smaller Cherry-faced Meadowhawk *(S. internum)* lacks stripes or spots on the thorax and has a broad, black lateral stripe on the abdomen. The Saffron-winged Meadowhawk *(S. costiferum)* also lacks thoracic stripes or spots and has a paler pterostigma. The Cardinal Meadowhawk *(S. illotum)* has dark brown or black patches in the wing bases and reddish legs. The Striped Meadowhawk *(S. pallipes)* has pale stripes on the front of the thorax. The Variegated Meadowhawk *(S. corruptum)* has a row of white spots on the sides of the abdomen.

BEHAVIOR: Females, in tandem or alone, lay eggs in water with emergent vegetation or on dry pond beds by dropping the eggs from the air. Nonbreeding individuals forage away from water in roadside bushes or the weedy, brushy borders of small clearings and stream banks. They typically forage from perches a few feet off the ground.

DISTRIBUTION: A somewhat local and relatively uncommon meadowhawk of northwestern North America, known in northern California southward to Santa Clara and Kern Counties, this species has also been found in southern California west of the deserts from Santa Barbara County to San Diego County (including the Channel Islands). Found from sea level to over 2,400 m (8,000 ft), it is undoubtedly more widely distributed in California than current records suggest.

HABITAT: The Red-veined Meadowhawk breeds at temporary and permanent ponds, lakes, sloughs, and streams. It seems to be more frequently found in riparian thickets and brushy areas near water than other meadowhawks, but it has also been found at hot springs and alkaline ponds and marshes in the Great Basin.

FLIGHT SEASON: It flies from April through September.

VARIEGATED MEADOWHAWK   *Sympetrum corruptum*
Pl. 31, Figs. 5, 13

LENGTH: 4 to 4.5 cm (1.5 to 2 in.); **WING SPAN:** 6 to 7.5 cm (2.5 to 3 in.)

DESCRIPTION: This is a medium-sized dragonfly with extremely variable coloration. Its most consistent and distinctive features are two diagonal, white stripes on the side of the thorax, each with a yellow spot at the lower tip (on the mature male only the yellow spots remain); white, black-rimmed spots (one per segment) along the sides of the abdomen, resembling a row of portholes; and a bicolored pterostigma that is dark in the center and pale yellow on either side. Coloration varies greatly with age, sex, and perhaps temperature. The portholes are most conspicuous on the female and the young male, when their black rims contrast with the pale dorsal surface (which is tan and white with orange bands) of the abdomen. These also have large, black spots atop abdominal segments 8 and 9, and the thorax is brown with white side and frontal stripes. The face is ivory to tan in color, the eyes above a light brown. The anterior wing veins are yellow to orange, but the membrane is washed with little if any tint. The legs are distinctly bicolored: black in front, tan behind. With age, the male develops pinkish red bands atop the abdomen which otherwise darkens to an olive gray, the pink color sometimes infusing the white spots and making them less conspicuous. The thorax ages to brown or gray, the white stripes reduced to two yellow dots low on the sides. The face and anterior wing veins become suffused with red, and the top of the eyes darkens to a rich reddish brown. Individuals seen in winter often seem drab and colorless, either because they are older or colder!

SIMILAR SPECIES: The pale spots that are low on the thorax and the pale anterior wing veins are features shared with other *"Tarnetrum"* meadowhawks (Red-veined and Cardinal *[S. madidum* and *S. illotum]*), and the Striped Meadowhawk *(S. pallipes)* has similar thoracic stripes, but the row of white spots on the abdomen is found only on the Variegated. The overall pinkish look of the mature male in flight is distinctive (this species has also been called the "pastel skimmer"), but drab gray or olive brown individuals seen at a distance can resemble other species, for example, gliders *(Pantala)* and the female Red Rock Skimmer *(Paltothemis lineatipes)*.

BEHAVIOR: They most frequently forage by sallying out from perches from ground level to the treetops but are also seen hawking at times over open areas or patrolling along shorelines and over ponds. Their flight is somewhat bouncy, with occasional sudden directional changes or quick stops to hover briefly. Pairs oviposit in tandem while flying over open water, the female tapping her abdomen on the surface a few times, then moving on a short distance to repeat the procedure. Migrations of Variegated Meadowhawks are only poorly understood. Apparently, individuals emerging in late winter and early spring in the south (Mexico, northward perhaps to the Central Valley) fly northward to breed in northern California and the Pacific Northwest. Individuals emerging in the north in summer and fall move southward, primarily in September and October, to breed in late fall in the southern part of the range. The appropriate autumnal weather conditions produce "fallouts" of migrants, much like those of migratory birds, which are particularly noticeable along the coast. Some perhaps survive through winter, especially in southern California.

DISTRIBUTION: Fairly common, even abundant in some areas, this is one of the most widespread and familiar dragonflies in California, found statewide from sea level to over 3,000 m (10,000 ft). Because of its migrations, it might be found anywhere, especially in fall. It is widely distributed throughout the western United States and southern Canada and is migratory in the East.

HABITAT: Its breeding habitat includes most freshwater environments, permanent and temporary. Nonbreeding individuals can be found away from water, typically in open habitats or woodland edge where they may appear in large numbers, but migrating individuals can show up anywhere, from urban yards to the open desert or even out at sea.

FLIGHT SEASON: This species has been recorded flying at all times of year. It is rarest in winter and most common in spring and fall.

CARDINAL MEADOWHAWK   *Sympetrum illotum*
Pl. 30, Fig. 13

LENGTH: 3 to 4 cm (1 to 1.5 in.); **WING SPAN:** 5.5 to 6 cm (2 to 2.5 in.)

DESCRIPTION: A very striking species, the male has the abdomen cardinal red without significant black markings, the thorax reddish brown with two white spots on each side. The abdomen is short and stout for a meadowhawk. The face and eyes are red, as are the anterior and basal wing veins. There are brown or black patches at the base of the wings in the male and female and a rusty wash (amber on the female) on the basal quarter of the wings, extending to the nodus along the costal stripe. The pterostigma is dark brownish. The immature male and the female are duller brown to rufous but have the characteristic wing pattern. The legs are rufous with black tarsi. The vulvar lamina resembles a small spout.

SIMILAR SPECIES: The Flame and Neon Skimmers *(Libellula saturata* and *L. croceipennis)* have similar color patterns but are much larger. The Red-veined Meadowhawk *(S. madidum)* has black legs and lacks dark patches at the wing bases.

BEHAVIOR: Breeding males are alert but rather unwary, sitting on exposed perches within one to a few feet above the ground or water, wings cocked forward, making short flights to feed, chase intruders, or patrol for females over the breeding pond or stream. Oviposition by females, usually in tandem while in flight, involves repeated dips of the abdomen tip into the water near shore. This species typically does not swarm in open meadows and fields as do other meadowhawks.

DISTRIBUTION: A widespread and familiar West Coast dragonfly, the Cardinal Meadowhawk is found statewide except for the Great Basin and Desert Provinces. It is quite common in the California Province, particularly the coastal districts, but relatively scarce in the more open countryside of the Central Valley. Primarily a foothill species, it has been collected as high as approximately 2,100 m (7,000 ft) in the Sierra Nevada. Its entire range extends from southern Canada to Argentina.

HABITAT: This species is found at small ponds and streams and the margins of larger lakes and rivers. It seems to prefer more wooded or swampy areas, including riparian strips and cattail marshes, than do other meadowhawks and is not often found foraging out in open country away from water.

FLIGHT SEASON: It typically flies from March to October, with individuals occasionally found in winter in southern California.

Tropical Pennants *(Brachymesia)*

Three species compose this small but diverse genus. One is found in the southeastern United States and Caribbean region, and two are rather widespread in the Neotropical region northward to the southern border of the United States. One of the latter species occurs in southern California. The genus's English name refers to its tropical affinities and habit of perching like small flags at the tips of upright sticks.

RED-TAILED PENNANT   *Brachymesia furcata*
Pl. 32

LENGTH: 4 to 4.5 cm (1.5 to 2 in.); **WING SPAN:** 6.5 to 7.5 cm (2.5 to 3 in.)

DESCRIPTION: The abdomen of the mature male is bright, scarlet red; otherwise this is a rather plain, medium-sized dragonfly. The mature male also has a reddish face, dark brown eyes, and a dark olive thorax. The legs are black but brown at their bases. The wings are mostly clear with a small wash of amber at the base of the hind wing. The pterostigma is brown. In profile, the male abdomen is distinctive: segments 2 and 3 are swollen, with a noticeable constriction between segments 3 and 4, and the superior appendages (cerci) curve sharply upward, like the tips of skis. The young male and most females lack red coloration; the abdomen is ochre or tan, and there are black oval spots atop segments 8 and 9.

SIMILAR SPECIES: It somewhat resembles a meadowhawk in general shape and color. The only meadowhawks in its range in California are the Variegated, Red-veined, and Cardinal *(Sympetrum corruptum, S. madidum,* and *S. illotum).* These meadowhawks have pale spots or stripes on the thorax and orange or red anterior wing veins, among other differences. The Roseate Skimmer *(Orthemis ferruginea)* is larger and has stripes on the sides of the thorax.

BEHAVIOR: Mature males perch conspicuously on twigs and sticks over open water, from which they forage and defend territories along the shoreline. Young males and females generally perch in less exposed situations, such as bushes and small trees. A female, the male hover-guarding nearby, releases eggs by dipping the tip of the abdomen in shallow water.

DISTRIBUTION: This southern species has been found along the Colorado River, near the Salton Sea, and at other scattered sites from Los Angeles and Riverside Counties to the Mexican border. Records are from locations below 300 m (1,000 ft).

HABITAT: Found at permanent ponds and lakes, it is capable of tolerating moderately polluted as well as brackish water.

FLIGHT SEASON: Dates in California range from May to November.

Pondhawks *(Erythemis)*

Pondhawks are a diverse genus of 11 species widely distributed in the Western Hemisphere. For their size, they are inclined to take fairly large prey, including other odonates.

WESTERN PONDHAWK   *Erythemis collocata*
Pl. 33

LENGTH: 4 cm (1.5 in.); **WING SPAN:** 6.5 cm (2.5 in.)

DESCRIPTION: This is a medium-sized dragonfly with clear wings. The face is green in both sexes and all ages. The female and the young male are almost entirely pale green, often yellower on the abdomen, which has a black stripe of varying thickness along the midline above. The mature male has the abdomen and thorax covered with blue pruinescence, and the eyes are dark blue above. The males takes a few weeks to change color; first the abdomen becomes blue, then the front of the thorax, and lastly, the sides of the thorax. The pterostigma is yellow to olive. The legs are mostly black, but pale yellow along the hind margin of the femur. The female has a spout-shaped ovipositor.

SIMILAR SPECIES: Some snaketails look mostly green and yellow at a distance but at close range are easily distinguished by their widely separated eyes, different abdominal patterns, and shape. The male Blue Dasher *(Pachydiplax longipennis)* is similar to the male pondhawk, and the two species are often found together, but the dasher is generally smaller, with a white face and green eyes. The male Bleached and Comanche Skimmers *(Libellula composita* and *L. comanche)* also have white faces and different pterostigma color. Darners *(Anax)* are much bigger, with longer abdomens, a different body shape overall, and much different behavior. The Western Pondhawk is sometimes considered conspecific with the Eastern Pondhawk *(E. simplicicollis),* which is unrecorded in California.

BEHAVIOR: These dragonflies perch on the ground, on surface vegetation, or low in dense vegetation bordering water or a clearing where, especially when green, they are difficult to spot. Even on breeding territories, males tend to perch either on floating vegetation or low stems near the water's edge. From these perch sites they target fairly large, low-flying prey, often taking bluets, forktails, and other damselflies. Females oviposit in flight by rapidly tapping the water with the tip of the abdomen a few times, then moving on to repeat the action at other spots in shallow water near shore. This is accomplished alone or with the male nearby flying tight, fast circles and chasing off intruders.

DISTRIBUTION: A common species in much of western North America from southern British Columbia south, it is found statewide in California except at higher elevations. It has been found from sea level to about 1,800 m (6,000 ft).

HABITAT: This species breeds in permanent and temporary waters, including springs, ponds, lakes, streams, river backwaters, and sloughs, especially with floating vegetation and algal mats. It is usually found near water, but nonbreeding individuals may wander some distance to forage in backyards and fields.

FLIGHT SEASON: It is on the wing from March through October.

Blue Dasher *(Pachydiplax)*

The sole member of this genus is a common and widespread dragonfly throughout much of the United States and parts of southern Canada.

BLUE DASHER *Pachydiplax longipennis*

Pl. 33

LENGTH: 3.5 to 4 cm (1.5 in.); **WING SPAN:** 5.5 to 6 cm (2 to 2.5 in.)

DESCRIPTION: This is a small to medium-sized dragonfly, the size varying with time of year (earlier emergers are larger). The scientific name refers to a thick *(pachys)* dragonfly with long wings *(longipennis)* and is particularly apt for the female, which has a relatively short and stout abdomen for its wing length. On the female and the young male, the thorax is striped with black or dark brown and yellow, and the abdomen is mostly black above with two parallel, broken yellow lines running down the length of the dorsal surface. The basal abdominal segments are orange below. On the mature male, the abdomen and often the thorax become a pruinose blue. The head of the mature male is distinctive, with the face ivory white and a small but brilliant, metallic blue green area on the frons between the emerald green eyes. The faces of of the female and the young male are similar to that of the mature male, but the eyes are brown above. The legs and terminal appendages are black. The wings are usually clear with small, black streaks at the very base, but sometimes with a light amber wash at the base, on the very tips, or as a vague cloud in midwing beyond the nodus.

SIMILAR SPECIES: The mature male Blue Dasher and mature male Western Pondhawk *(Erythemis collocata)* are similar and often found together. The latter is generally larger and has a green face. The larger Comanche Skimmer *(Libellula comanche)* male has a bicolored pterostigma. The Bleached Skimmer *(L. composita)* is also larger and has pale eyes and costa. The female's body pattern is unlike that of other similarly sized species such as meadowhawks, although at a distance this is hard to see.

BEHAVIOR: Blue Dashers actively sally for small prey, returning repeatedly to the same perch after making a capture. They look very alert on perch, seeming to lean forward with the abdomen cocked up and the wings swept forward or up and back. Males occupy low perches over or near water and, where densities are high, engage in frequent territorial skirmishes for preferred sites. Females typically stay some distance from the water except for reproductive purposes and are often found foraging from fairly high perches (3 to 6 m [10 to 20 ft]) in small trees. Females oviposit in flight, alone or guarded by a hovering male, by making occasional downward taps with the tip of the abdomen to wash eggs off at the water's surface.

DISTRIBUTION: Widely distributed and common in California except in the northeast corner, this is typically a species of low to moderate elevations, unrecorded above 2,000 m (6,500 ft).

HABITAT: Frequenting a wide range of permanent and temporary sites, including ponds, marshes, and lakes. it is also found at sloughs and the slow backwaters and stagnant pools of river channels and streambeds. The female and the young male forage away from water in backyards, woodland edges, and clearings with small, scattered trees.

FLIGHT SEASON: Records span nearly the entire year, from February to November.

Amberwings *(Perithemis)*

The 12 species in this distinctive genus of little, yellow dragonflies are primarily Neotropical in distribution. Of the three species found in the United States, one is common and widespread in the East. The other two occur in the Southwest, one of them in California.

MEXICAN AMBERWING   *Perithemis intensa*
Pl. 32

LENGTH: 2.5 cm (1 in.); **WING SPAN:** 4 to 4.5 cm (1.5 to 2 in.)

DESCRIPTION: This is a dumpy, little, yellow dragonfly. The abdomen is short and spindle shaped, the wings relatively short and broad. The male, in particular, is butterscotch yellow nearly throughout, including the wings. The eyes are brown, and there are some faint, dark streaks atop the abdominal segments. The pterostigma is reddish orange. The female is similar but often duller, with a different wing pattern that is variable but usually has some patches of amber and/or brown at and behind the nodus and in the area between the nodus and the wing base.

SIMILAR SPECIES: The Mexican Amberwing is probably more likely to be mistaken for a large wasp than other species of dragonfly. The Cherry-faced Meadowhawk *(Sympetrum internum)* can be plain bodied with some amber wash in the wing, but it has a different wing pattern, thin abdomen, and doesn't overlap in range or habitats occupied.

BEHAVIOR: These are perch-and-sally feeders that sit, pennantlike, at the tips of twigs with the wings and abdomen often raised. Males are territorial, actively patrolling from low perches over water in defense of oviposition sites (plants, woody debris, etc., at water level). In their courtship flight, males lead females to these sites, where the females place eggs near the water's edge. Nonbreeding individuals occupy exposed perches away from water, sometimes a few meters (6 to 12 ft) above the ground in brush and small trees.

DISTRIBUTION: Most common within California in the Imperial Valley and along the Colorado River, this species has also been collected in northern Inyo County and westward to the Los Angeles area. Recorded locations have typically been within 300 m (1,000 ft) above or below sea level, but the site in Inyo County

(Eureka Valley) is at about 900 m (3,000 ft). It is also found in Arizona and New Mexico southward into Mexico.

HABITAT: The Mexican Amberwing is found around the often stagnant margins of ponds, lakes, ditches, and river lagoons.

FLIGHT SEASON: This species can be found in flight from April to September.

Whitetails *(Plathemis)*

Whitetails get their name from the white pruinescence that covers the abdomens of mature males. Both species in the genus are North American, and both occur in California. They bear a resemblance to some black-and-white king skimmers *(Libellula)* and have often been placed in that genus, but recent molecular genetic data suggest they are a distinct group.

COMMON WHITETAIL   *Plathemis lydia*
Pls. 34, 35, Fig. 4

LENGTH: 4.25 to 4.75 cm (1.5 to 2 in.); **WING SPAN:** 6.5 to 7.5 cm (2.5 to 3 in.)

DESCRIPTION: A stout dragonfly with a relatively broad and stubby abdomen, the mature male is strikingly marked with a chalky white, pruinescent abdomen and a thick, black band on each wing just beyond the nodus. There are also black basal stripes in each wing, those of the hind wings bordering small triangles of white pruinescence at the wing base (occasional pruinescence occurs at the base of the fore wings, too). The head, eyes, and thorax are brown. The top of the thorax, around the wing bases, is often pruinose white. The immature male is similar but lacks pruinescence and has the abdomen brown with creamy white spots forming a distinctly broken "chain" along each side above. A complex pattern on the side of the thorax (brighter on the female and the young male, obscured on the mature male) includes a wavy, often interrupted, anterior white stripe with a disconnected yellow dot below it; and a shorter, posterior, white stripe with a yellow dot included in its lower tip. The body of the female is similar to that of the immature male, but the wing pattern is very different, with three dark brown to black patches on each wing: a basal stripe like the male's, a blotch of irregular size and shape at midwing around the nodus, and a dark wing tip beyond the pterostigma (which is also black).

SIMILAR SPECIES: The male Desert Whitetail *(P. subornata)* is similar, but the mature male has extensive white pruinescence on the basal half of the wings, and the dark wing bands often appear to have three parts (dark on the sides and paler in the middle) and lack a triangular projection along the costal stripe extending from the nodus toward the wing base, which is present on the Common Whitetail. The female Desert Whitetail lacks dark wing tips, but the female Twelve-spotted Skimmer *(Libellula pulchella)* has a nearly identical wing pattern. The Twelve-spot has a longer, slimmer abdomen with yellow side stripes, versus the broken chain of spots on the female Common Whitetail's abdomen.

BEHAVIOR: Rather unwary dragonflies, whitetails perch on or near the ground and fly low—from about knee-high to chest-high—with a smooth, sailing flight. Males favor conspicuous, sunny spots and defend shoreline territories with an abdomen-lifting display and frequent patrolling. Females perch on dense vegetation in small, semishady clearings in nearby woods, where they are difficult to spot. The female deposits eggs by repetitiously slapping shallow water near the shore with the tip of the abdomen, while the male hovers nearby to ward off intruders.

DISTRIBUTION: One of the most familiar and widely distributed dragonflies in the United States, the Common Whitetail occurs throughout most of California as well, although more commonly in the northern half of the state than south of the Tehachapi Mountains. The only part of California that lacks records is the southeastern desert quarter, from Mono County southward to the Arizona and Mexican borders. Records range from sea level to 2,000 m (6,500 ft).

HABITAT: This whitetail frequents a diversity of aquatic habitats: ponds, marshes, lakes, roadside ditches, streams, sloughs, and the backwater lagoons of large rivers. More or less permanent water with a muddy bottom seems to be preferred. This species can tolerate polluted waters, so it is often found at ponds in urban parks.

FLIGHT SEASON: It flies from March to October.

DESERT WHITETAIL   *Plathemis subornata*

Pl. 34, 35

LENGTH: 4 to 5 cm (1.5 to 2 in.); **WING SPAN:** 6.5 to 7.5 cm (2.5 to 3 in.)

DESCRIPTION: This is a small, stocky dragonfly. The mature male has the abdomen and basal half of the wing pruinose white, and a black to brown band, often darkest on the sides and lighter down the middle, on each wing just beyond the nodus. A small, black stripe extends out from each wing base. The head, eyes, and thorax are brown. The female and the immature male lack white pruinescence and have two rows of large, yellow ovals forming broken lateral stripes atop the brown abdomen. The thorax has two yellow lateral stripes on each side and a pair of yellow frontal stripes. Each wing has a black stripe at its base and two wavy, black to brown bands, one at the nodus and one at the anterior end of the pterostigma, corresponding to the dark sides of the mature male's wing band. The female has a yellowish face and large, yellow spots on the abdomen below. The pterostigma is dark brown to black.

SIMILAR SPECIES: The male Common Whitetail *(P. lydia)* lacks extensive white pruinescence on the basal half of the wings. The female Common Whitetail has dark wing tips and a broken chain of white spots on the sides of the abdomen. The Eight-spotted Skimmer *(Libellula forensis)* is larger, with a dark wing patch not in contact with the pterostigma and, in the male, a white pruinescent patch near the pterostigma.

BEHAVIOR: Mature males are quick, agile fliers, patrolling and defending territories along spring runs and the edges of spring pools. Aerial chases are common. They fly just above the surface of the ground or water, perching briefly on low vegetation nearby. A little distance from water, females and nonbreeding males perch on the low edges of sagebrush *(Artemisia)*, rabbitbrush *(Chrysothamnus)*, and other desert shrubs. They are hard to spot until flushed and easy to lose track of as they flit through the brush.

DISTRIBUTION: This species is found at scattered sites in the southern third of California, from Los Angeles, Kern, and San Bernardino Counties southward and along the eastern border of the state in the Great Basin Province in Modoc, Lassen, and Inyo Counties. Occupied locations range from sea level to 1,500 m (5,000 ft).

HABITAT: In general, small wetland sites in arid landscapes are frequented, particularly spring runs, seep pools, and small, spring-fed streams. It often occurs at hot springs. Aquatic vegetation at such sites typically consists of a relatively narrow band of dense sedges, grasses, and other low herbaceous growth.

FLIGHT SEASON: Dates in California range from April to September.

Corporals (Ladona)

This genus is closely related to the king skimmers (Libellula) and is often placed with them. The four or five species of corporals are of Holarctic distribution, with three species in North America, one of them found in California.

CHALK-FRONTED CORPORAL　　　　　　　　*Ladona julia*
Pl. 26

LENGTH: 4 to 4.5 cm (1.5 to 2 in.); **WING SPAN:** 6 to 7 cm (2.5 to 3 in.)

DESCRIPTION: The mature male has two parallel, white bars on the front of the thorax (the "corporal's stripes" for which this species is named) and bright white pruinescence on the basal third of the abdomen. The mature female may resemble the male; otherwise, it is a rather nondescript, midsized, dark gray brown dragonfly. The female has a short, thick abdomen with the basal abdominal segments paler gray above. The immature female is paler and browner. The wings are mostly clear except for small, black, roughly triangular patches on the hind wing bases.

SIMILAR SPECIES: The pattern of white on the adult male, coupled with the mostly clear wings, is distinctive. The nondescript female and the young male, however, might be mistaken for the Four-spotted Skimmer (Libellula quadrimaculata) or a baskettail (Tetragoneuria). The Four-spotted Skimmer has nodal wing spots and a pale face. Baskettails have butterscotch yellow spots on the sides of the middle abdominal segments and hang from perches in vegetation, rarely landing on the ground.

BEHAVIOR: Males patrol territories along the lakeshore, where they frequently perch on logs, rocks, or the shore itself. Away from water, foraging individuals may perch low on tree trunks. They fly low over the water or ground and are rather unwary.

DISTRIBUTION: This species occupies wooded lakes across Canada and the northern United States. In California it is known from a handful of locations in the northern third of the state (Fish Lake, Humboldt County; Castle and Gumboot Lakes, Siskiyou County; the Lassen Peak region; McMurray Lake, Nevada County; and Miller Lake, Placer County). Except for Fish Lake, at 600 m (1,900 ft), occupied lakes range from 1,500 to 2,100 m (5,000 to 7,000 ft) in elevation. The Sierra Nevada populations are the southernmost known in western North America.

HABITAT: The Chalk-fronted Corporal typically occurs at mountain lakes containing floating vegetation, such as pond-lilies *(Nuphar)*.

FLIGHT SEASON: Its known season in California—from the first week of June to the first week of August—is short and, especially for a montane species, somewhat early.

King Skimmers *(Libellula)*

This large, diverse genus of 24 species contains some of the most familiar dragonflies in the Northern Hemisphere. Most (18 species) occur in North America, nine of these in California. Two or more species may often be found around ponds, lakes, marshes, springs, sloughs, wet meadows, or slow backwaters of rivers and streams in just about any part of the state during spring and summer months. They typically perch in exposed situations, from which they sally out or hawk for brief periods to capture prey. Many are rendered all the more conspicuous, and readily identified, by their brightly colored abdomens and distinctive wing patches.

Four of our species have wings that are strikingly spotted with black or black and white in males. Males of these species develop white or gray pruinose areas on the abdomen as well as on the wings when mature, whereas females and young males typically show yellowish spots or stripes on the abdomen. Some bear a close resemblance to the whitetails *(Plathemis)*. Take note of both the wing pattern and the abdominal pattern—spotted or striped—to identify these species. Two species, the Flame and Neon Skimmers *(L. saturata* and *L. croceipennis)*, are large and robust and have red or rust-colored bodies with washes of red on the wing membranes. They can be confused with Roseate and Red Rock Skimmers *(Orthemis ferruginea* and *Paltothemis lineatipes)*.

Among king skimmers, mating is brief, often completed in flight. Females then oviposit by dipping the abdomen in water or splashing (most with the aide of small flaps on the sides of abdominal segment 8) eggs mixed with water drops up onto the shoreline. Females typically oviposit alone or with males hovering nearby (tandem oviposition is documented for a few species).

FOUR-SPOTTED SKIMMER   *Libellula quadrimaculata*
Pl. 26

LENGTH: 4.5 cm (2 in.); **WING SPAN:** 6.5 to 7 cm (2.5 to 3 in.)

DESCRIPTION: This is a medium-sized, rather drab skimmer. A black, triangular patch at the base of the hind wing, pale-veined and bordered in front with an amber stripe, was counted as a spot when the species was named; the four spots are the small, black spots on each nodus. The pterostigma is black or dark brown. The pale yellow face contrasts with the dark brown eyes. The hairy thorax is pale olive with some fine, black scrawling on the sides. The tapered abdomen above looks as if dipped in ink; it is olive colored on the basal segments and black from the middle of segment 6 to the tip, with narrow, pale yellow lateral stripes. The mature male may have some gray pruinescence on the middle abdominal segments. The female resembles the male, and the immature individual is more brightly marked in tan and yellow tones.

SIMILAR SPECIES: Other king skimmers with small, dark spots on the nodus include some Hoary and Bleached Skimmers *(L. nodisticta* and *L. composita)*. The Hoary Skimmer has a black stripe near the base of the fore wing and a dingy face. The Bleached Skimmer has a white costa and pale eyes and thorax and lacks the dark triangular patch at the base of the hind wing. The Chalk-fronted Corporal *(Ladona julia)* lacks nodal wing spots. Baskettails *(Tetragoneuria)* also lack spots at the nodus and have a spindle-shaped abdomen.

BEHAVIOR: Males at breeding sites occupy low, exposed perches on emergent vegetation or sticks along the shoreline or out over water. From there they sally out to feed, chase other males, or pursue females. The females oviposit, with the male hover-guarding nearby, by dipping the abdomen into water while in flight or perched.

DISTRIBUTION: This is a common and familiar species in Europe and Asia as well as in much of North America. In California, it is common at higher elevations, from 1,200 to 2,700 m (4,000 to 9,000 ft), in the North Coast Ranges southward to Mendocino and Lake Counties, the Cascade-Sierran Province southward to Tulare County, and the Great Basin Province. It is typically the most abundant libellulid at mountain ponds and lakes in summer. A couple of puzzling records from southern California (Orange and San Diego Counties) suggest it might occur in some southern mountain ranges.

HABITAT: This is a species of marshes, ponds, lakes, springs, and bogs, but it is also found at slow backwaters of mountain streams.

FLIGHT SEASON: This species flies from mid-May through mid-September, with most activity from June through August.

TWELVE-SPOTTED SKIMMER   *Libellula pulchella*
Pls. 34, 35

LENGTH: 5.5 cm (2 in.); **WING SPAN:** 8.5 to 9 cm (3.5 in.)

DESCRIPTION: This is a sleek skimmer with the "trim" of a fighter plane. The 12 black spots (three on each wing at the base, nodus, and tip) are concentrated along the leading halves of the wings, making the wings appear narrower than those of other similar skimmers. The mature male has white pruinescent spots (four per wing) between the black spots. Both sexes have pale yellow, lateral stripes on the abdomen and two pale yellow, diagonal slashes on each side of the dark thorax. The abdominal pattern of the mature male is obscured by whitish or bluish gray pruinescence.

SIMILAR SPECIES: The Twelve-spotted Skimmer can be told from the Eight-spotted Skimmer *(L. forensis)* by simply counting the black spots on the wings. However, some teneral or worn Twelve-spots may have only faint gray or brown washes representing the spots on the wing tips. In these cases, note the midwing spots, which are larger and nearly in contact with the trailing wing edge on the Eight-spot (see the discussion in the "Similar Species" section of the Eight-spotted Skimmer account). A Twelve-spot with faded wing tips also resembles the Hoary Skimmer *(L. nodisticta),* but the latter species has much-reduced spots at the nodus and pale spots rather than a continuous stripe on the sides of the abdomen. The female Common Whitetail *(Plathemis lydia)* has a similar wing pattern of 12 black spots, and the two species are often found together. The Common Whitetail has lateral rows of pale, cream-colored spots on the abdomen rather than the continuous pale stripes on the Twelve-spot; it is also dumpier, with a short, thick abdomen.

BEHAVIOR: Their breeding behavior is much like that of other black-and-white skimmers. Foraging away from water, they perch at moderate heights (1.5 to 6 m [5 to 20 ft]) in exposed situations. When they perch, the wings are usually held out straight and stiff, not appearing to "droop" as much as those of the Eight-spotted and Widow Skimmers *(L. luctuosa).*

DISTRIBUTION: This is a widespread and familiar skimmer throughout much of North America. In California, it is found throughout the northern two-thirds of the state and, apparently in small numbers, south in the coastal counties of Los Angeles, Orange, and San Diego at elevations ranging from sea level to at least 2,100 m (7,000 ft). There are no records for the southern deserts.

HABITAT: The Twelve-spotted Skimmer is found near marshes, springs, ponds, lakes, rivers, and creeks. It often can be found foraging away from water in yards and grassy fields.

FLIGHT SEASON: This species is active from April through October, with the largest numbers found May to July.

EIGHT-SPOTTED SKIMMER   *Libellula forensis*
Pls. 34, 35
LENGTH: 5 cm (2 in.); **WING SPAN:** 7.5 to 8 cm (3 in.)

DESCRIPTION: This is a medium-sized skimmer with a distinctive, dark, blackish or brown blotch shaped like a figure eight at midwing; the wing tips are clear. These midwing blotches, plus dark blotches on each wing base, give the species its eight "spots." In the mature male, the abdomen and front of the thorax are thinly pruinose, light gray to white, and there are patches of white pruinescence on the wings—one at each hind wing base, between the basal and midwing dark patches and behind each pterostigma. Most females and the young male lack pruinescence and have two lateral rows of yellow ovals on the abdomen, which form broken stripes. Some old females become pruinose like the male. There are two yellow stripes on the side of the brown thorax, the anterior one usually interrupted in the middle.

SIMILAR SPECIES: The Eight-spotted Skimmer is distinguished from the Twelve-spotted Skimmer *(L. pulchella)* by the lack of dark wing tips. The wing tips of some female Eight-spots may have a smoky wash, however; and beware of teneral or worn Twelve-spots, which may show only a gray, ghostly trace of the wing-tip pattern. The basal and midwing dark blotches (and the white blotches on the wings of the male) are more extensive, especially toward the rear margins of the wings, than are the comparable spots on the wings of the Twelve-spot. This difference makes the wings of the Eight-spot seem bigger and wider. The Eight-spot in general seems more robust and less sleek than the Twelve-spot. The Desert Whitetail *(Plathemis subornata)* is superficially similar, and its range in the Great Basin broadly overlaps that of the Eight-spot. The outermost dark wing band on the Desert Whitetail comes in contact with the pterostigma (isolated in the Eight-spot), and the male lacks white pruinescence in the outer half of the wings. The female Desert Whitetail is smaller and has stockier proportions and two midwing bands. The Hoary Skimmer *(L. nodisticta)* has a much smaller dark spot around the nodus.

BEHAVIOR: Their mating behavior is like that of the Widow Skimmer *(L. luctuosa)* and other *Libellula* species. Foraging individuals disperse over marshes, fields, and meadows. They are typical perch-and-sally feeders.

DISTRIBUTION: The Eight-spot is widespread and locally common in the northern two-thirds of California southward to Santa Cruz, Kern, and Inyo Counties. There is but a single, old record in southern California (Orange County). It can be particularly abundant, outnumbering all other skimmers, in marshes of the Great Basin Province in midsummer but is scarce on many parts of the Central Valley floor, especially where the Widow Skimmer *(L. luctuosa)* is common. Restricted to western North America, this species is also known as the Western Widow Skimmer.

HABITAT: This species primarily inhabits marsh-bordered ponds, as well as slow-moving rivers and creeks bordered with extensive emergent vegetation.

FLIGHT SEASON: It is on the wing from April through September, with a peak in the summer months (June and July).

HOARY SKIMMER   *Libellula nodisticta*
Pls. 34, 35
LENGTH: 4.5 to 5 cm (2 in.); **WING SPAN:** 7.5 to 8 cm (3 in.)
DESCRIPTION: A robust, medium-sized skimmer, its most distinctive features are on the wings; each wing has a short, thick, black stripe at the base and a small, dark patch at the nodus. The dark nodal patch is, in general, larger on specimens from west of the Pacific Crest and smaller in Great Basin populations. The pterostigma is black. Mostly brown bodied, the female and the young male exhibit two yellow, interrupted stripes (which

look like four oval spots) on each side of the thorax and a dorso-lateral row of yellow spots on abdominal segments 2 through 8. The mature male develops a gray pruinescence on the thorax and abdomen, obscuring the yellow marks, and some white pruines-cence on the wings surrounding the black basal stripes. The eyes are dark brown, the face dingy, with a yellow frons on the female and a black frons on the male.

SIMILAR SPECIES: The Bleached Skimmer *(L. composita)* can have a nodal spot as large or larger than some Great Basin Hoary Skimmers, and both species occur at hot springs in the region. The Bleached Skimmer has pale sides on the thorax, has a pale face and eyes, and lacks heavy, dark, basal wing stripes. The Four-spotted Skimmer *(L. quadrimaculata)* has a pale face and a dark patch at base of the hind wing only. The Eight-spotted and Twelve-spotted Skimmers *(L. forensis* and *L. pulchella)* have larger dark patches at midwing.

BEHAVIOR: They are often found away from water, perched in brush and low trees bordering open areas or in open woodland. Females oviposit by slapping the tip of the abdomen at the water's surface, usually alone, but occasionally in tandem.

DISTRIBUTION: Although found widely within California from Ore-gon to the Mexican border, this skimmer has a patchy distribution apparently influenced by the presence or absence of springs. Areas of occurrence include foothill and lower mountain zones of the Cascade-Sierran Province southward to Mariposa County and along the North Coast Ranges southward to Napa and Sonoma Counties, the Great Basin Province from Modoc County south-ward to Inyo County, and the foothills and mountains of the south-ern coastal slope from Los Angeles and San Bernardino Counties southward to San Diego County. There are few if any records from the Central Valley, Desert Province, or central Coast Ranges. Recorded elevations range from near sea level to 2,000 m (6,700 ft).

HABITAT: This species is consistently associated with springs and spring-fed streams, including hot springs. Because such breed-ing sites are typically localized and scattered through a variety of landscapes, from conifer forests to sagebrush deserts, non-breeding individuals can be found foraging in a wide range of nearby habitats, typically clearings and meadows in forested country, foothill oak woodlands, and shrub-dominated com-munities.

FLIGHT SEASON: It has been recorded flying from late April to mid-September.

WIDOW SKIMMER *Libellula luctuosa*

Pls. 34, 35

LENGTH: 4.5 to 5 cm (2 in.); **WING SPAN:** 7.5 to 8 cm (3 in.)

DESCRIPTION: This is a medium-sized skimmer with a broad, dark brown to black band covering the basal third of the wings. This band may appear nearly solid in color or paler (but still cloudy or dingy brown) in the center and darker along the outer margin. The mature male is pruinose white at midwing bordering the dark basal band, as well as on the abdomen and front of the thorax. The eyes and face are dark. The female and the young male are dark bodied with a yellow stripe on top of the thorax between the wing bases that forks at the base of the abdomen to form two yellow, dorsolateral stripes along its length. The female's wing pattern is like that of male, but often with a dark smudge at the wing tips (occasionally on the male, too).

SIMILAR SPECIES: Once familiar, the Widow Skimmer is not easily mistaken for anything else. Saddlebags *(Tramea)* also have dark, basal wing bands but differ in proportions and lack the extensive white areas on the wings and bodies of the mature male Widow Skimmer or the yellow abdominal stripes of the female and the young male. Other black-and-white skimmers have distinctly different wing patterns, all differing from the Widow Skimmer in lacking dark coloring along the entire width of the hind wing base. The wing patterns can be faint on teneral individuals, however, so caution is urged in identifying them. The Western Meadowhawk *(Sympetrum occidentale)* has brown wing bases but is much smaller and lighter (yellowish or red) in body color.

BEHAVIOR: Mature males usually stick close to water, where they sit on emergent vegetation above the surface and maintain territories, darting out occasionally to chase other males or pursue females. Their flight low over the ground or water often seems a bit erratic or choppy, almost butterflylike. Females and young males may be seen far from water foraging in weedy fields or yards. They often sit low in weeds or grass with their wings drooped but may occasionally be seen perched in treetops.

DISTRIBUTION: The Widow Skimmer is found throughout most of the United States and is widely distributed in southern California and west of the Pacific Crest in northern California, as far north as Humboldt and Siskiyou Counties. It is relatively scarce in the montane zones of the Sierra Nevada, Cascade Range, or Coast

Ranges, where the species is occasionally found around lakes to (rarely) as high as 1,700 m (5,500 ft).

HABITAT: This skimmer is often associated with artificial bodies of water or waters heavily influenced (e.g., polluted) by human activity. Apparently not known in the north of the state in the early 1900s, this species has expanded its range with human modification of the landscape and waterways. Individuals may be found along the quiet backwaters of rivers and streams, as well as along the vegetated shores of ponds and lakes. This species seems particularly fond of areas with floating vegetation, such as waterweeds (*Ludwigia*).

FLIGHT SEASON: The flight season broadly overlaps that of other black-and-white-winged skimmers, but the peak of the season in general seems to be a bit later, with good numbers not seen until June and July, and moderate numbers still present into October.

COMANCHE SKIMMER   *Libellula comanche*
Pl. 36

LENGTH: 4.75 to 5.5 cm (2 in.); **WING SPAN:** 7 to 9 cm (3 to 3.5 in.)

DESCRIPTION: The mature male of this handsome skimmer has a white face, pearly blue gray eyes, and powder blue pruinescence on the thorax and abdomen above. The sides of the thorax are a pale, dingy white with one dark, central stripe. The wing membranes are essentially clear with perhaps a light amber wash on the costal stripe, powder blue pruinescence on the base of the hind wing, and a thin, smoky rim extending from the pterostigma around the wing tip. This is our only skimmer with a distinctly bicolored pterostigma: it is mostly bright white and a small amount of the outer tip is black. The female and the young male have the thorax broadly striped in brown and white, and the abdomen is dark brown to black above with broad, pale yellow, dorsolateral stripes. The pterostigma is like that of the mature male, bicolored black and (mostly) white, the white areas sometimes tinged brown on the female. The wing tips of the female are edged with a narrow, brown wash. The eyes are light brown to gray.

SIMILAR SPECIES: The striking white pterostigma distinguishes it from all our other skimmers, including the Bleached Skimmer (*L. composita*), which is also pale bodied and is found at some of the same desert springs occupied by the Comanche Skimmer.

BEHAVIOR: They perch on vegetation near water, often rather high atop weed stalks, cattails, bushes, and even small trees. Males at breeding sites aggressively chase other males, even harassing dragonflies of other species.

DISTRIBUTION: This is a relatively rare skimmer in California, known from Modoc and Inyo Counties in the Great Basin Province, southward into San Bernardino, Riverside, San Diego, and Imperial Counties in the Desert Province at elevations ranging from below sea level (Salton Sea vicinity) to 1,400 m (4,600 ft). On the Pacific slope, it has been found at only a handful of foothill springs north to Napa and Butte Counties. It is widely distributed in the American Southwest and northern Mexico.

HABITAT: This species is typically found at springs (including hot springs) and spring-fed pools and streams with dense emergent vegetation, often including cattails *(Typha)*. Most occupied sites are small wetlands in arid, brushy landscapes.

FLIGHT SEASON: California dates are from May through September.

BLEACHED SKIMMER   *Libellula composita*
Pl. 36

LENGTH: 4 to 5 cm (1.5 to 2 in.); **WING SPAN:** 7.5 to 8 cm (3 in.)

DESCRIPTION: The mature male is a pale apparition on the wing. The face is mostly white, the eyes a pearly gray. The thorax looks bleached, pruinescent white on the sides to powder blue above with blue lines along some sutures. The abdomen is also pruinescent pale blue. The white or pale yellow costa is a distinctive feature of both sexes. The pterostigma is black. There is often a small, dark brown spot at the nodus and an amber wash and/or thin brown streaks on the wing base. The adult female has a pale face and eyes like the male. Younger individuals of both sexes may have brown eyes above. The young male and the female have a mostly white thorax striped with brown and black; the abdomen is black above with two broken, pale yellow, dorsolateral stripes.

SIMILAR SPECIES: The Comanche Skimmer *(L. comanche)* looks similar on the wing or at a distance, but it can be distinguished easily if the mostly white pterostigma is seen. The Hoary Skimmer *(L. nodisticta)* has a dark black stripe on each wing base, darker eyes and face, and dark costa. The Blue Dasher *(Pachydiplax longipennis)* is smaller with dark eyes and costa.

BEHAVIOR: Bleached Skimmers are often seen on the wing near water and not easily found perched. They perch low on weeds and brush. Oviposition is done in tandem (which is fairly unusual for a king skimmer), the females tapping the abdomen at various spots on the water surface.

DISTRIBUTION: Local in distribution and seldom encountered, this ghostly desert species is known from a few sites in the Great Basin (Modoc, Mono, and Inyo Counties) and the southern deserts (Imperial and Riverside Counties) at elevations ranging from below sea level near the Salton Sea to over 1,800 m (6,000 ft) in the Mono Basin.

HABITAT: Breeding sites are small, shallow, alkaline pools. The species is often found at springs, including hot springs.

FLIGHT SEASON: It has been seen on the wing from May into September.

NEON SKIMMER   *Libellula croceipennis*
Pl. 37

LENGTH: 5.5 cm (2 in.); **WING SPAN:** 8 to 9.5 cm (3 to 4 in.)

DESCRIPTION: The mature male is a big skimmer that is red nearly throughout and brilliant "neon" red on the abdomen. The face is red, the eyes brown. The wing (including veins) is washed with rust color on the basal quarter, extending to the nodus only along the costal stripe. The pterostigma is orange red and about 6 mm long. The female is tawny in color where the male is red, the thorax has a white, central stripe on the front and between the wing bases, and there is little color on the wings, reduced primarily to rusty veins on the anterior of the wing as far as the nodus. The female has broad flaps on the sides of abdominal segment 8. The immature male is colored more like the female than the mature male.

SIMILAR SPECIES: The Flame Skimmer *(L. saturata)* is quite similar, being distinguished by brown stripes on the wing bases and a more extensive wash of rust in the wing on the male. The Cardinal Meadowhawk *(Sympetrum illotum)* is smaller, with pale spots on the side of the thorax and dark patches in the wing bases. The male Red Rock Skimmer *(Paltothemis lineatipes)* has extensive black markings on the thorax and abdomen. The female Roseate Skimmer *(Orthemis ferruginea)* has stripes on the sides of the thorax.

BEHAVIOR: They often perch on brush, brambles, and small trees near breeding sites. Males frequently engage in aerial skirmishes. Their breeding behavior is similar to that of Flame Skimmers.

DISTRIBUTION: This is primarily a middle American dragonfly (ranging south to Colombia) that just enters the American Southwest. In California, it occurs in the Desert Province, on the southern coastal plain northward to Los Angeles County, and at scattered locations to the north, including Antelope Spring, Inyo County, and a few primarily foothill sites from Kern County to Shasta County, the latter including the northernmost records for the species. Elevations of record range from sea level to about 1,700 m (5,500 ft).

HABITAT: This species is found at spring runs and along small streams bordered by riparian vegetation or open woods. Occupied sites often have very shallow water and extensive subsurface vegetation.

FLIGHT SEASON: The available records suggest that it flies late, from June to mid-October.

FLAME SKIMMER   *Libellula saturata*
Pl. 37

LENGTH: 5 to 6 cm (2 to 2.5 in.); **WING SPAN:** 8.5 to 9.5 cm (3.5 to 4 in.)

DESCRIPTION: The mature male is bright orange red nearly throughout (brown eyes and thorax). The wings are washed with orange rust on the basal half to the nodus; the anterior wing veins are orange and yellow. Each wing has a brown stripe in the basal quarter. The pterostigma is orange red and about 5 mm long. The immature male and the female have a tawny abdomen and a white stripe on top of the light brown thorax (as on the Neon Skimmer *[L. croceipennis]*). The female's wing is colored like that of the male along the costal stripe to the pterostigma (yellow and orange veins, washed with rust) but lacks the rust wash on the posterior basal half. The flaps on abdominal segment 8 are not as broad as on the female Neon Skimmer.

SIMILAR SPECIES: The Cardinal and Red-veined *(Sympetrum illotum* and *S. madidum)* meadowhawks are much smaller and have spots or stripes on the thorax. The male Red Rock Skimmer *(Paltothemis lineatipes)* has extensive black markings on the thorax and abdomen. The female Roseate Skimmer *(Orthemis ferruginea)* has stripes on the thorax and a dark pterostigma. The Neon Skimmer is the most similar species (see the "Description" section for that species). The best distinguishing characteristic in both sexes of the Flame Skimmer is the brown, basal wing stripe, which is lacking in the Neon Skimmer, but this feature is also absent on some female Flame Skimmers. In such cases, look for a more extensive wash of amber or rust on the costal stripe on the Flame Skimmer and, in hand, a shorter pterostigma (less than 5.5 mm; greater than 5.5 mm on the Neon Skimmer).

BEHAVIOR: Nonbreeding individuals foraging away from water are primarily perch-and-sally feeders, perching in exposed situations from within a foot or two of the ground to the tops of trees. Mature males at breeding sites occupy perches at medium heights on shoreline vegetation from which they make frequent, brief flights to hawk for food over water, patrol in search of females or intruders, or chase other males. Mating occurs briefly in flight, after which the females oviposit by using the tip of the abdomen to flip drops of water, carrying eggs, upto the shore at or near the water line. They often do this alone, at various spots around the pond or along the stream pool, but are sometimes guarded by the male, hovering nearby.

DISTRIBUTION: A common and familiar skimmer in the American West, this species is found nearly throughout California below 2,000 m (6,500 ft). It is most typical of arid valley and foothill sites. There are few records for the humid northwest corner of the state.

HABITAT: It is found along slow streams, irrigation ditches, river backwaters, marshy ponds, and springs. Nonbreeding individuals wander to suburban yards, parks, and open woodland.

FLIGHT SEASON: This skimmer has been reported flying in all months of the year in southern California.

Tropical King Skimmers *(Orthemis)*

As the English name implies, this genus of 18 or so species replaces the similar king skimmers *(Libellula),* to some extent, in the Neotropics. Two species reach the southern United States, one of them in southern California.

ROSEATE SKIMMER   *Orthemis ferruginea*
Pl. 32

LENGTH: 5.5 cm (2 in.); **WING SPAN:** 8 to 9.5 cm (3 to 3.5 in.)

DESCRIPTION: This is a good-sized dragonfly. The mature male has a distinctive plum-colored pruinescence on the thorax and rose pink or violet pruinescence on the abdomen. The frons is metallic purple above, the eyes a rich, purple brown. A narrow, dusky brown stripe extends along the leading edge of the wing around the tip from the end of the black pterostigma. The wings are otherwise clear. A red-colored male is known from elsewhere in its range, but it has not been found in California. The female and the young male have a brown thorax, striped and mottled on the sides with white. Their abdomens are rust colored above, striped black and white below. The eyes and frons are brown. A white stripe extends from the top front of the thorax between the wing bases and onto the top of abdominal segments 1 and 2. The female has lateral flaps on abdominal segment 8, as on king skimmers *(Libellula)*.

SIMILAR SPECIES: None of our other dragonflies is colored like the mature male. The female and the young male Neon and Flame Skimmers *(Libellula croceipennis* and *L. saturata)* are similar in size and behavior and have similar abdominal color, but they lack stripes on the side of the thorax and have pale orange pterostigma and anterior wing veins. The Red-tailed Pennant *(Brachymesia furcata)* is smaller and has a plain, olive brown thorax.

BEHAVIOR: Tropical king skimmers behave much like king skimmers. Away from water, they are primarily perch-and-sally feeders, sitting on exposed twigs at the tops of bushes and trees. At breeding ponds, some males occupy low perches over water in defended territories or make short patrols along the shore in search of ovipositing females. Other males perch a short distance away and attempt to intercept arriving females. Copulation is brief and achieved in flight. The females then oviposit while hovering, using the abdomen to slap water mixed with eggs onto the shore or emergent vegetation at the water line, while the male hover-guards nearby.

DISTRIBUTION: This species is fairly common from the Arizona and Mexican borders northward in the Desert Province to Riverside County, with a few records from the southern coastal plain northward to Los Angeles County, at elevations below 700 m (2,200 ft). It ranges southward to Costa Rica.

HABITAT: This species breeds in stagnant and temporary ponds, lakes, ditches, and riparian backwaters, tolerating brackish or polluted conditions. It forages away from water in brushy areas, parks, yards, and other open areas with scattered trees or brush.

FLIGHT SEASON: Although scarce in winter, it has been found flying throughout the year in southern California.

Rock Skimmers *(Paltothemis)*

There are two or three middle American species in this genus, one reaching the southwestern United States, including California. In appearance and behavior, they seem somewhat intermediate between typical skimmers and the broad-winged gliders.

RED ROCK SKIMMER   *Paltothemis lineatipes*
Pl. 38, Fig. 8

LENGTH: 4.5 to 5.5 cm (2 in.); **WING SPAN:** 9 to 9.5 cm (3.5 in.)

DESCRIPTION: The mature male is mottled rufous and black on the side of the thorax and the top of the abdomen. The face and the front of the thorax are red, the eyes are red brown. The basal third of each wing is rusty (including veins and tinted membrane). Viewed from below (as when hawking insects overhead), the body of the male looks black. The female and the young male are gray where the mature male is red (so they look rather nondescript), are mottled black and light gray, have clear wings without rusty tints, and have brown eyes. The pterostigma is dark brown.

SIMILAR SPECIES: Flame and Neon Skimmers *(Libellula saturata* and *L. croceipennis)* have robust, red abdomens lacking extensive black markings. Cardinal and Red-veined Meadowhawks *(Sympetrum illotum* and *S. madidum)* are smaller and have white spots on the thorax and different wing patterns. Gliders *(Pantala)* somewhat resemble the female Red Rock Skimmer but have brown or orange abdomens rather than gray, and the Spot-winged Glider *(P. hymenaea)* has a dark hind wing spot.

BEHAVIOR: At streams, territorial males sit on sunny rocks near defended oviposition sites or patrol a low, steady path at moderate speed with occasional gliding flight along stretches of the streambed, looking for females. Males usually hover-guard ovipositing females, but where male densities are low along intermittent streams, unattended females may make rather quick, furtive visits to oviposit by tapping the tip of the abdomen in pools in the streambed. When foraging away from water, both sexes often fly rather high, 3 to 6 m (10 to 20 ft) up, and hawk back and forth along a regular path, much like gliders or saddlebags *(Tramea)*.

DISTRIBUTION: Within the state this species is more or less restricted to the foothills of the California Province from Mendocino and Shasta Counties southward to the Mexican border. It occurs in some desert canyons on the west side of the Desert Province and up to about 1,500 m (5,000 ft) in some mountain canyons but is unrecorded in either the high Sierra Nevada or the Great Basin.

HABITAT: Breeding occurs along rocky stream courses in arid scrub, foothill oak woodland, or conifer forests, but this species can be found hawking for food over brush, open woods, and trails away from water.

FLIGHT SEASON: It is on the wing in California from late March to early November.

Clubskimmers *(Brechmorhoga)*

These skimmers resemble the members of the clubtail family (Gomphidae) in appearance and preferred habitat, hence the English name. The 14 species in the genus are primarily Neotropical, but two occur in the southwestern United States, one of them reaching California.

PALE-FACED CLUBSKIMMER   *Brechmorhoga mendax*

Pl. 38, Fig. 8

LENGTH: 5.25 to 6.25 cm (2 to 2.5 in.); **WING SPAN:** 7.5 to 9 cm (3 to 3.5 in.)

DESCRIPTION: The stout thorax and slender, black abdomen ending in a "club" (expanded abdominal segments 7 through 9) on the male of this large, gray, clear-winged dragonfly give it the appearance of a clubtail (Gomphidae). The most distinctive feature that catches the eye, especially on a flying Pale-faced Clubskimmer, is a large white spot (which is actually a pair of spots separated by a thin black line) on segment 7. The face is a pale, dull yellow, and the eyes are dark, gray brown and come in contact atop the head. The thorax is gray with pale lateral stripes (dull white with a slight green tint). The sides of the basal abdominal segments have white patches (the next-most visible pale area, after the white spot on the club, on a flying individual), and some small streaks and spots of white on the dorsal abdominal segments are visible only at close range on the perched Pale-faced Clubskimmer. The female resembles the male in pattern, but the abdomen is less slender and not clubbed at the tip, and the wing tips and a small area at the wing bases are washed with light brown.

SIMILAR SPECIES: This species most closely resembles clubtails, rather than any other skimmer, in flight. Clubtails fly low over streams but often stop to perch on rocks, the ground, or low vegetation along the shore. The eyes of clubtails do not touch each other.

BEHAVIOR: These have been described as among the most graceful of our dragonflies. In late morning, males cruise regular beats low over streams, foraging, waiting for females, and chasing intruders. They seem in continual motion, never stopping to sit on sunny rocks as a clubtail would. A whirling pair of males engaged in a territorial skirmish move so quickly, they become a fuzzy blur in which individual shapes are lost. When not at breeding sites, both males and females wander away to feed in hawking flight, often flying regular routes at heights of 3 to 6 m [10 to 20 ft] over weedy clearings, meadows, and trails. When they hang up to perch, they do so in inconspicuous spots on weeds, shrubs, or up in trees.

DISTRIBUTION: A dragonfly of lower elevations (to about 900 m [3,000 ft]), it is found in the Desert Province and the valleys and foothills of the California Province northward to Shasta County. It ranges southward into Mexico.

HABITAT: Clear streams and shallow rivers with a mix of riffles and gravel-bottomed pools are the haunts of this species. Nonbreeding individuals may be found some distance from water foraging in open woodland, yards, and weedy clearings.

FLIGHT SEASON: This species is on the wing from April to October.

Marl Pennants *(Macrodiplax)*

There are only two species in this genus, one widely distributed in southern parts of the Eastern Hemisphere from Africa to Japan and Australia, the other found in the Western Hemisphere from the southern United States to Venezuela. They are typically associated with brackish or marl-lined ponds.

MARL PENNANT *Macrodiplax balteata*
Pl. 39
LENGTH: 4 cm (1.5 in.); **WING SPAN:** 7 cm (3 in.)
DESCRIPTION: This is a small, alert dragonfly with relatively broad hind wings and a short abdomen. It has long legs for its size. The mature male appears mostly black in the field. The eyes are a rich

maroon brown to almost black above, and there is a small but contrasty white spot behind each eye on the sides of the head. Both sexes have an oval, black-brown patch on the anterior base of the hind wing. In profile, the male's abdomen is somewhat expanded at the base and a little at the tip; it is carried with a slight arch in flight. The female and the young male have a white face, brown eyes above, and a pale, olive gray thorax with black and brown stripes that form a vague W on the side. Abdominal segments 1 through 7 are dull yellow edged with black, including a black midline stripe, and segments 8 through 10 are black. The pterostigma is short and yellow.

SIMILAR SPECIES: Saddlebags *(Tramea)* are larger and have a more extensive black patch on the hind wing base. The Spotwinged Glider *(Pantala hymenaea)* has a smaller black wing spot near the trailing edge of the wing base. The female Comanche and Bleached Skimmers *(Libellula comanche* and *L. composita)* are larger, have paler eyes, and lack the patch on the hind wing base.

BEHAVIOR: Mature males patrol fast and low over ponds, stopping often to hover briefly. Away from water they perch, flaglike, from exposed twigs of bushes and trees, with the abdomen and wings frequently raised. Oviposition occurs in tandem.

DISTRIBUTION: The Marl Pennant is of local occurrence in the Desert Province (Riverside and Imperial Counties) in the southeastern corner of the state. Elevations of occurrence are at or near (many are below) sea level. It has probably spread into the southeastern corner of the state within the last century as a result of human manipulation of the Colorado River, including the inadvertent creation of the Salton Sea.

HABITAT: A species of brackish waters and mineral spring pools, it may also be found along irrigation ditches and other artificial bodies of water.

FLIGHT SEASON: It has been reported flying from May through September in California.

Saddlebags *(Tramea)*

The English name for these dragonflies comes from the dark patches at the base of the hind wings, which supposedly resemble a horse's saddlebags. This is a large genus of impressive dragonflies, with some 24 species found nearly worldwide. Of the seven species occurring in North America, two are found in

California. Many species of saddlebags are far-flung wanderers, however, so additional species may yet be recorded in the state. A good candidate would be the Striped Saddlebags *(Tramea calverti)*, which has been found in Arizona and Baja California. Oviposition in this genus is unique: flying low in tandem over water, the male quickly releases the female, who drops to the surface and taps a cluster of eggs off the tip of the abdomen, then instantly the male recaptures her to fly on and repeat the process.

BLACK SADDLEBAGS   *Tramea lacerata*
Pl. 39
LENGTH: 5 to 5.5 cm (2 in.); **WING SPAN:** 9.5 to 10 cm (3.5 to 4 in.)
DESCRIPTION: This is a large, sleek, nearly all black dragonfly. Large, black, irregularly shaped patches on the hind wing bases resemble ink blots. The sexes are similar in appearance. The immature form is slightly paler, with purple or brown tones and large, paired dorsal spots of pale yellow on abdominal segments 3 through 7, these darkening with age; those on segment 7 often persist, especially on the female.
SIMILAR SPECIES: The Red Saddlebags *(T. onusta)* on the wing looks similar at a distance. At close range, notice the red wing veins, which give the saddle a red sheen, and the red face and abdomen. The Marl Pennant *(Macrodiplax balteata)* is smaller, with a smaller hind wing patch. The Widow Skimmer *(Libellula luctuosa)* has an abdomen either striped with yellow or pruinose white.
BEHAVIOR: Males patrol widely over breeding ponds with swift, low flight. The abdomen is carried straight or with a slight droop in hawking flight. Away from water, the Black Saddlebags forages from 3 to 30 m (10 to 100 ft) up by flying along regular beats for extended periods. They usually hang up to roost high in trees but are occasionally found perching horizontally on tall weeds in open country. They often swarm with other species, especially gliders *(Pantala)* and Common Green Darners *(Anax junius)*. Their seasonal movements are not precisely understood, but there seems to be movement of individuals—probably having emerged to the south and at lower elevations—to the north and higher elevations, apparently to breed there and then die. A cor-

responding southward and downslope movement in fall presumably represents their offspring, returning to continue the cycle.

DISTRIBUTION: This is a strong-flying species that has been recorded throughout the state. Most records are at lower elevations, but it has been found up to about 1,700 m (5,500 ft) in the Sierra Nevada. Apparently some seasonal movements occur in spring and fall.

HABITAT: It is typically found at ponds, lakes, and ditches, often temporary ones, but it also breeds in still, backwater lagoons of rivers and marshy sloughs. Nonbreeding individuals forage widely away from water over grasslands, meadows, suburban yards, open woodland—almost anywhere.

FLIGHT SEASON: This species is seen April through October.

RED SADDLEBAGS   *Tramea onusta*
Pl. 39

LENGTH: 4 to 5 cm (1.5 to 2 in.); **WING SPAN:** 8 to 9 cm (3 to 3.5 in.)

DESCRIPTION: This dragonfly is similar to the Black Saddlebags *(T. lacerata)* in overall shape and pattern, but its face and abdomen are dull red, with black patches atop segments 8 and 9, and its thorax is olive brown and unpatterned. The eyes are red brown. The basal hind wing patches have ragged edges and are dark brown with red veins, so they look dark red in flight. The female and the young male are similar to the adult male, but the red areas are a duller, tawny red.

SIMILAR SPECIES: The Black Saddlebags is black, often with a pale yellow abdominal spot, and has black saddlebags. The Striped Saddlebags *(T. calverti),* unrecorded in California, has pale stripes on the side of the thorax and smaller saddlebags with smooth outer margins.

BEHAVIOR: They behave similarly to Black Saddlebags. The abdomen often seems to have a more pronounced droop in flight than shown by Black Saddlebags. They seem constantly on the wing over marsh-bordered ponds and sloughs. Females oviposit alone as well as in tandem.

DISTRIBUTION: This is a southern species found in the Desert Province, along the Pacific slope from Los Angeles County southward, and the Great Basin from Mono County southward. It is usually found near sea level but has been noted to about 1,700 m (5,500 ft) in the Great Basin. A handful of recent records from northern California, from Sonoma County southward and mostly along the coast, suggest that it is at least an occasional wanderer there.

HABITAT: Like the Black Saddlebags, it is found at marshy ponds, lakes, and ditches.

FLIGHT SEASON: California dates range from April to October.

Gliders *(Pantala)*

The two widely distributed species in this genus have relatively short, streamlined bodies and broad-based hind wings that are perfect for nearly effortless, gliding flight. They can stay on the wing for hours, even days, when making transoceanic flights. The center of their distribution is tropical, but a large number fly north to the temperate zone to breed in summer. They move on prevailing air currents and often appear in numbers following the passage of monsoon rains. After mating, females oviposit in flight by making a series of abdominal taps at the water surface. They are also known as rainpool gliders. Larval growth is quick, to accommodate the often ephemeral nature of breeding ponds, and the summer's offspring make the return flight south to breed in autumn and winter.

SPOT-WINGED GLIDER *Pantala hymenaea*
Pl. 40

LENGTH: 4.5 to 5 cm (2 in.); **WING SPAN:** 8.5 to 9.5 cm (3.5 in.)

DESCRIPTION: This species has a typical glider profile, slim and broad winged. The wings are clear except for a round brown spot at the posterior base of the hind wings. Veins in this spot become lighter with age, and the spot may be obscured. The pterostigma is tan. The face is yellow, but red on the mature male. The eyes are red brown above, gray below. The thorax is olive gray, paler on sides with a faint indication of brown stripes sometimes present. The legs are long and mostly black except for a pale, straw-colored base and a rear stripe on the femur. The abdomen is tawny above, whiter below, and intricately patterned as follows: striped with black below, these lines are wavy on the lower sides of segments 2 through 4; fine, black rings (one or two per segment, along the dorsal carinae) on segments 1 through 5; paired, white oval patches on segments 3 through 8 separated by a dark, dorsal stripe that forms black, anchor-shaped marks on segments 4 through 8, and black ovals atop segments 9 and 10. This pattern is brightest on the young Spot-winged Glider and fades to dull gray and tan with age.

SIMILAR SPECIES: Saddlebags *(Tramea)* have larger wing patches and are black or red on the abdomen. The Wandering Glider *(P. flavescens)* lacks the dark wing spot, but this is hard to see on many Spot-winged Gliders in flight. The black spot on the wing base of the smaller Marl Pennant *(Macrodiplax balteata)* is at the leading edge of the wing next to the thorax; the Spot-winged Glider's wing spots are adjacent to the abdomen.

BEHAVIOR: They are often seen on the wing, from 2 to 30 m (6 to 100 ft) up, in seemingly constant search for prey. They feed on small, swarming insects; for example, in the lee of eucalyptus trees infested with lerp psyllids (small aphidlike insects [Psyllidae: Homoptera]). They migrate and forage in swarms of up to 100 or more individuals. These swarms may be mixed-species groups containing Wandering Gliders, saddlebags, Common Green Darners *(Anax junius),* and occasionally other species. Gliding flight is interrupted by frequent, quick darts to catch prey. Their abdomens often seem to droop slightly in flight. When coming to roost, they quickly approach and then back away from a tree or bush a number of times before hanging up at a perch. A typical perch site is near the tip of a hanging branch, from which they hang like a tree ornament. The tip of the abdomen is sometimes curled up when perched. At high densities, they may roost in groups. They prefer perches over 2 m (6 ft) above the ground in trees, large bushes, or vines but will perch in low brush in open country.

DISTRIBUTION: This species is found nearly statewide, usually in modest numbers, occasionally in abundance. There are no certain records for the northwestern corner of the state, but it probably occurs there and has been overlooked. Its wind-borne wanderings have taken it as high as 3,300 m (11,000 ft) in the Sierra Nevada, but it is typically a species of lower elevations. The Spot-winged Glider is widely distributed in the Western Hemisphere from southern Canada to Chile.

HABITAT: This glider breeds in shallow, open water with relatively sparse vegetation, including rain pools, still backwaters in stream- and riverbeds, rice fields, sewage ponds, and similar sites. It can tolerate brackish or polluted waters. Rapid larval growth allows use of ephemeral ponds. Nonbreeding individuals hawk incessantly over many terrestrial habitats, even open desert.

FLIGHT SEASON: It is found flying from March through October and is most numerous in midsummer, June through August.

WANDERING GLIDER   *Pantala flavescens*

Pl. 40

LENGTH: 4.5 to 5 cm (2 in.); **WING SPAN:** 8 to 9 cm (3 to 3.5 in.)

DESCRIPTION: This is a slim, broad-winged glider with mostly clear wings lacking dark spots or patches. The wing tips and the base of the hind wing may have an amber wash of variable extent and intensity. The pterostigma and the undersides of the anterior wing veins are golden. Otherwise, it is patterned much like the Spot-winged Glider *(P. hymenaea)* except that the pale areas of the abdomen above tend to be yellow, becoming bright yellow or-ange on the mature male.

SIMILAR SPECIES: Although unmistakable when seen perched at close range, this species is easily confused in flight with the Spot-winged Glider. Caution is advised, as the wing spots on the latter species are sometimes so obscure that they are difficult to see on a flying individual. The Variegated Meadowhawk *(Sympetrum corruptum)* in sustained flight might be mistaken for this species, but it has white spots on the side of the abdomen and eventually comes to perch in a more or less horizontal position.

BEHAVIOR: Wandering Gliders behave much like Spot-winged Gliders. They often occur with that species in feeding swarms and at roost sites, but they show a greater tendency to perch low (within 1 m [3 ft] of the ground) in weedy areas.

DISTRIBUTION: This is a true wanderer, found nearly worldwide except in Europe and Antarctica, including on many oceanic is-lands (e.g., Hawaii). In California, it has been recorded through-out southern California, the Central Valley, and the Great Basin Province. There are a few records for the central Coast Ranges, and there appear to be a few reports from above 1,200 m (4,000 ft). It is likely to show up anywhere, though, and will probably be found in all parts of the state eventually.

HABITAT: For breeding, temporary ponds, rain pools, and rice fields are used. Nonbreeding individuals, especially on migration, may be found, like the Spot-winged Glider, nearly anywhere.

FLIGHT SEASON: It is on the wing from May to November, broadly overlapping the season of the Spot-winged Glider, but peaking somewhat later in fall (August to October).

CHECKLIST OF CALIFORNIA DRAGONFLIES AND DAMSELFLIES

Damselflies (Zygoptera)

BROAD-WINGED DAMSELS (CALOPTERYGIDAE)

☐ River Jewelwing *(Calopteryx aequabilis)*
☐ American Rubyspot *(Hetaerina americana)*

SPREADWINGS (LESTIDAE)

☐ California Spreadwing *(Archilestes californica)*
☐ Great Spreadwing *(Archilestes grandis)*
☐ Spotted Spreadwing *(Lestes congener)*
☐ Common Spreadwing *(Lestes disjunctus)*
☐ Emerald Spreadwing *(Lestes dryas)*
☐ Black Spreadwing *(Lestes stultus)*
☐ Lyre-tipped Spreadwing *(Lestes unguiculatus)*

POND DAMSELS (COENAGRIONIDAE)

☐ California Dancer *(Argia agrioides)*
☐ Paiute Dancer *(Argia alberta)*
☐ Emma's Dancer *(Argia emma)*
☐ Lavender Dancer *(Argia hinei)*
☐ Kiowa Dancer *(Argia immunda)*
☐ Sooty Dancer *(Argia lugens)*
☐ Powdered Dancer *(Argia moesta)*
☐ Aztec Dancer *(Argia nahuana)*
☐ Blue-ringed Dancer *(Argia sedula)*

- ☐ Vivid Dancer *(Argia vivida)*
- ☐ Taiga Bluet *(Coenagrion resolutum)*
- ☐ River Bluet *(Enallagma anna)*
- ☐ Double-striped Bluet *(Enallagma basidens)*
- ☐ Boreal Bluet *(Enallagma boreale)*
- ☐ Tule Bluet *(Enallagma carunculatum)*
- ☐ Familiar Bluet *(Enallagma civile)*
- ☐ Alkali Bluet *(Enallagma clausum)*
- ☐ Northern Bluet *(Enallagma cyathigerum)*
- ☐ Arroyo Bluet *(Enallagma praevarum)*
- ☐ Exclamation Damsel *(Zoniagrion exclamationis)*
- ☐ Desert Forktail *(Ischnura barberi)*
- ☐ Pacific Forktail *(Ischnura cervula)*
- ☐ Black-fronted Forktail *(Ischnura denticollis)*
- ☐ Swift Forktail *(Ischnura erratica)*
- ☐ San Francisco Forktail *(Ischnura gemina)*
- ☐ Citrine Forktail *(Ischnura hastata)*
- ☐ Western Forktail *(Ischnura perparva)*
- ☐ Rambur's Forktail *(Ischnura ramburii)*
- ☐ Sedge Sprite *(Nehalennia irene)*
- ☐ Desert Firetail *(Telebasis salva)*
- ☐ Western Red Damsel *(Amphiagrion abbreviatum)*

Typical Dragonflies (Anisoptera)

PETALTAILS (PETALURIDAE)

- ☐ Black Petaltail *(Tanypteryx hageni)*

DARNERS (AESHNIDAE)

- ☐ Common Green Darner *(Anax junius)*
- ☐ Giant Darner *(Anax walsinghami)*
- ☐ California Darner *(Aeshna californica)*
- ☐ Canada Darner *(Aeshna canadensis)*
- ☐ Variable Darner *(Aeshna interrupta)*
- ☐ Blue-eyed Darner *(Aeshna multicolor)*
- ☐ Paddle-tailed Darner *(Aeshna palmata)*
- ☐ Shadow Darner *(Aeshna umbrosa)*

☐ Walker's Darner *(Aeshna walkeri)*
☐ Riffle Darner *(Oplonaeschna armata)*

CLUBTAILS (GOMPHIDAE)

☐ Grappletail *(Octogomphus specularis)*
☐ Pacific Clubtail *(Gomphus kurilis)*
☐ Brimstone Clubtail *(Stylurus intricatus)*
☐ Olive Clubtail *(Stylurus olivaceus)*
☐ Russet-tipped Clubtail *(Stylurus plagiatus)*
☐ Bison Snaketail *(Ophiogomphus bison)*
☐ Great Basin Snaketail *(Ophiogomphus morrisoni)*
☐ Sinuous Snaketail *(Ophiogomphus occidentis)*
☐ Pale Snaketail *(Ophiogomphus severus)*
☐ White-belted Ringtail *(Erpetogomphus compositus)*
☐ Serpent Ringtail *(Erpetogomphus lampropeltis)*
☐ Gray Sanddragon *(Progomphus borealis)*

SPIKETAILS (CORDULEGASTRIDAE)

☐ Pacific Spiketail *(Cordulegaster dorsalis)*

CRUISERS, EMERALDS AND BASKETTAILS, AND SKIMMERS (LIBELLULIDAE)

Cruisers (Macromiinae)

☐ Western River Cruiser *(Macromia magnifica)*

Emeralds and Baskettails (Cordulinae)

☐ American Emerald *(Cordulia shurtleffii)*
☐ Ringed Emerald *(Somatochlora albicincta)*
☐ Mountain Emerald *(Somatochlora semicircularis)*
☐ Beaverpond Baskettail *(Tetragoneuria canis)*
☐ Spiny Baskettail *(Tetragoneuria spinigera)*

Skimmers (Libellulinae)

☐ Crimson-ringed Whiteface *(Leucorrhinia glacialis)*
☐ Hudsonian Whiteface *(Leucorrhinia hudsonica)*
☐ Dot-tailed Whiteface *(Leucorrhinia intacta)*
☐ Red-waisted Whiteface *(Leucorrhinia proxima)*

- [ ] Variegated Meadowhawk *(Sympetrum corruptum)*
- [ ] Saffron-winged Meadowhawk *(Sympetrum costiferum)*
- [ ] Black Meadowhawk *(Sympetrum danae)*
- [ ] Cardinal Meadowhawk *(Sympetrum illotum)*
- [ ] Cherry-faced Meadowhawk *(Sympetrum internum)*
- [ ] Red-veined Meadowhawk *(Sympetrum madidum)*
- [ ] White-faced Meadowhawk *(Sympetrum obtrusum)*
- [ ] Western Meadowhawk *(Sympetrum occidentale)*
- [ ] Striped Meadowhawk *(Sympetrum pallipes)*
- [ ] Yellow-legged Meadowhawk *(Sympetrum vicinum)*
- [ ] Red-tailed Pennant *(Brachymesia furcata)*
- [ ] Western Pondhawk *(Erythemis collocata)*
- [ ] Blue Dasher *(Pachydiplax longipennis)*
- [ ] Mexican Amberwing *(Perithemis intensa)*
- [ ] Common Whitetail *(Plathemis lydia)*
- [ ] Desert Whitetail *(Plathemis subornata)*
- [ ] Chalk-fronted Corporal *(Ladona julia)*
- [ ] Comanche Skimmer *(Libellula comanche)*
- [ ] Bleached Skimmer *(Libellula composita)*
- [ ] Neon Skimmer *(Libellula croceipennis)*
- [ ] Eight-spotted Skimmer *(Libellula forensis)*
- [ ] Widow Skimmer *(Libellula luctuosa)*
- [ ] Hoary Skimmer *(Libellula nodisticta)*
- [ ] Twelve-spotted Skimmer *(Libellula pulchella)*
- [ ] Four-spotted Skimmer *(Libellula quadrimaculata)*
- [ ] Flame Skimmer *(Libellula saturata)*
- [ ] Roseate Skimmer *(Orthemis ferruginea)*
- [ ] Red Rock Skimmer *(Paltothemis lineatipes)*
- [ ] Pale-faced Clubskimmer *(Brechmorhoga mendax)*
- [ ] Marl Pennant *(Macrodiplax balteata)*
- [ ] Black Saddlebags *(Tramea lacerata)*
- [ ] Red Saddlebags *(Tramea onusta)*
- [ ] Wandering Glider *(Pantala flavescens)*
- [ ] Spot-winged Glider *(Pantala hymenaea)*

SPECIES OF HYPOTHETICAL OCCURRENCE

A number of species not included in the main list of this book have been reported for California based on published accounts or labeled specimens in collections. Some of these reports are relatively easily discounted because they involve species that are otherwise unknown from any other locations close to the state's boundaries. Examples include the Ebony Jewelwing *(Calopteryx maculata)* and Eastern Forktail *(Ischnura verticalis)*. Even if specimens of these species were actually collected in California, they probably represent ephemeral and inadvertent introductions from elsewhere.

However, there are reports for California of species found in neighboring states that, although not entirely convincing, raise the possibility that these species occur naturally within our boundaries. Some are discussed below. For identification of these species, consult the "References" section.

Canyon Rubyspot *(Hetaerina vulnerata):* Specimens have been reported collected east of Atwater, Merced County. Recent efforts to find this species there have been unsuccessful. The nearest known locations of occurrence are in the canyonlands of southern Nevada and northern Arizona. The Atwater location is far from the known range and has no suitable habitat, suggesting the record refers to either mislabeled specimens or the introduction of an ephemeral population. However, the species occurs close to the Nevada border and may yet be found in the eastern Mojave

Desert. This species closely resembles the American Rubyspot *(H. americana)* and can safely be identified only in the hand.

Lance-tipped Darner *(Aeshna constricta):* Old records of this species for California are the result of confusion with other similar darners (e.g., Paddle-tailed Darner *[A. palmata])* before species limits and ranges of the mosaic darners were well defined; however, it has been recorded in Nevada and Oregon and might conceivably occur in extreme northwest California.

Sedge Darner *(Aeshna juncea):* As in the case of the Lance-tipped Darner, an old record for California probably involves confusion with some other species of mosaic darner, but the range of this species in Oregon suggests it might be found in California.

Black-winged Dragonlet *(Erythrodiplax funerea):* An old specimen from northern California (Santa Clara County) of this tropical species is probably mislabeled, but it has strayed to Arizona and might occasionally reach southeastern California.

Filigree Skimmer *(Pseudoleon superbus):* This strikingly marked species occurs in Baja California and Arizona, but old references in the literature of occurrence in California are unconfirmed. Like the Black-winged Dragonlet, however, which it resembles, it could reach our southern border on occasion.

Striped Saddlebags *(Tramea calverti):* This saddlebags is of regular occurrence in south Texas and also occurs in Baja California. It has a tendency to wander in late summer and autumn, and there are numerous records throughout the eastern United States. Hence, it might be expected to wander northward to California occasionally.

GLOSSARY

Abdomen The hindmost body division, typically long and thin in odonates and having 10 segments.

Abdominal appendages Also known as *caudal appendages,* or *terminal appendages.* These extend from the tip of the last (tenth) abdominal segment and include the cerci, or superior appendages (two in all males and females), and inferior appendages (two para-procts in male zygopterans, a single epiproct in male anisopter-ans). They are modified in males for grasping females (*see* cercus, epiproct, paraproct).

Anal loop A group of cells near the base of the hind wing that varies in shape among families of Anisoptera.

Andromorphic (coloration) Female coloration similar to that of males.

Carina A narrow keel or ridge on a body plate. The plural is *carinae.*

Cell A membranous area of the wing bordered by veins.

Cercus One of a pair of upper abdominal appendages, also known as *superior appendages* (*see* abdominal appendages). The plural is *cerci.*

Clypeus The central plate or region of the face.

Costa The wing vein that forms the leading edge of the wing.

Costal stripe The area of the wing immediately behind the costa.

Emergence Here used to refer to the release of the adult form from the final larval instar (*see* instar, molt).

Epiproct In male anisopterans, the inferior abdominal appendage. In male dancers *(Argia)*, it is a small lobe projecting from the top center of the tenth abdominal segment, between the two cerci.

Exuvia The shed exoskeleton of a larval odonate. The plural is *exuviae.*

Femur The large segment of the leg near its base.

Frons The top plate or region of the face, sometimes called the *forehead.*

Genitalia (secondary) The structures on the underside of abdominal segment 2 and the front part of segment 3 of males and used to transfer sperm to females in the wheel (copulatory) position (*see* hamules).

Gills The three leaf-shaped terminal appendages on damselfly larvae. Also called *caudal laminae.*

Gynomorphic (coloration) Female coloration different from that of males.

Hamules Paired, hooklike structures on either side of the penis, which are part of the secondary genitalia.

Instar The larval stage between molts. The final instar immediately precedes emergence of the adult form.

Labial mask Formed by the labial palps and the front of the prementum. It covers the lower part of the face of some larval dragonflies (*see* labial palp, prementum).

Labial palp One of a pair of hinged flaps, usually with hooks or hairs, at the front of the prementum, used in prey capture.

Labium As used here, the hinged, extendible lower "jaw" of larval odonates, consisting of postmentum, prementum, and labial palps.

Larva The early, aquatic life stage(s) of odonates. The plural is *larvae.*

Mesostigmal plates A pair of small body plates at the top front edge of the pterothorax, just behind (and often partly covered by) the prothorax. Variation in shape of these plates is useful in distinguishing the females of some damselflies.

Molt The shedding of the larval exoskeleton (*see* exuvia, instar).

Nodus The thickened area of veins at the midpoint of the leading edge of each wing.

Obelisk position The position adopted by a perched dragonfly, with the tip of the abdomen pointed toward the sun (typically, when the sun is overhead) in order to reduce exposed surface area and prevent overheating.

Occipital horns The paired horns atop the occiput, useful in identification of some female clubtails.

Occiput The upper rear surface of the head between the compound eyes.

Oviposition The placing of eggs by the female.

Ovipositor The well-developed structure on the underside of abdominal segment 9 in some species of odonates (all damselflies, some anisopterans), used to insert eggs into a substrate.

Paraproct One of a pair of inferior appendages projecting from the lower tip of the abdomen in male damselflies, used, in concert with the cerci, to grasp females in tandem (*see* cercus).

Postmentum The hind portion of the larval labium, folded backward at rest, and thrust forward to extend the prementum (*see* prementum).

Postoccipital horns Paired horns on the rear margin of the occiput, useful in the identification of some female clubtails.

Prementum The front part of the labium, hinged to the postmentum and held below the body, then thrust forward in prey capture. Usually shaped something like a paddle or spoon (*see* labial palps, postmentum).

Prothorax The first thoracic segment (between the head and pterothorax). Usually small and inconspicuous, it bears the front pair of legs.

Pruinose (body) Coated with a white, gray, or pale bluish powdery substance (pruinescense), produced in maturity on the surfaces of certain body parts in some odonates.

Pterostigma A thickened, usually colored wing cell near the tip of the wing along the leading edge (just behind the costa), present in most odonates.

Pseudopupil A dark spot in the compound eye, resembling a pupil and indicating an area of acute vision.

Pterothorax The large middle section of the body, formed by fusion of the second and third thoracic segments, it houses the large flight muscles and bears the wings and middle and hind pairs of legs. In this guide, it is usually simply called the *thorax.*

Radial planate An arcing wing vein running parallel with, and posterior to, the radial sector in the middle of the outer half of the wing. Whether a dragonfly has one or two rows of cells between the radial planate and the radial sector is of help in distinguishing some species.

Radial sector A major vein in the middle of the outer wing.

Spurs Sharp, spiny projections on the legs.

Stylus One of a pair of thin projections at the tip of the ovipositor. The plural is *styli.*

Tandem The position in which the male holds the female by her head or thorax with his abdominal appendages.

Tarsus The end of the leg (the "foot"), consisting of three small segments and tipped with two claws. The plural is *tarsi.*

Teneral The adult form within a few hours after emergence, still soft bodied and relatively colorless.

Tibia The middle segment of the leg, between the femur and the tarsus. The plural is *tibiae.*

Torus One of a pair of thickened pads along the upper edge of the tenth abdominal segment above the cerci of male dancers *(Argia).* The plural is *tori.*

T-spot A dark, T-shaped patch on the top of the frons of some species *(see* frons).

Triangle A triangular cell or group of cells near the base of the wing.

Tubercle A bump or knob on the surface of the body.

Vertex The area on the top front of the head between the compound eyes, bearing three simple eyes.

Vulvar lamina The plate on the lower surface of abdominal segment 9 of females of some species, used in oviposition (*see* ovipositor). The plural is *vulvar laminae.*

Vulvar spine A backward projecting spine at the rear base of abdominal segment 8 in some species of damselflies.

Wheel position A copulatory position in which the male's terminal appendages attach to the female's head or thorax, and the tip of the female's abdomen attaches to the male's secondary genitalia, forming a closed circle, or wheel.

REFERENCES

Argia is the quarterly newsletter of the Dragonfly Society of the Americas. Membership in the society is $15 per year. The society also publishes a journal, *Bulletin of American Odonatology.* For more information write to DSA, c/o T. Donnelly, 2091 Partridge Lane, Binghamton, NY 13903.

Biggs, K. 2000. *Common dragonflies of California: A beginner's pocket guide.* Sebastopol, Calif.: Azalea Creek Publishing. Has excellent photographs of the more common and conspicuous species.

Corbet, P.S. 1963. *A biology of dragonflies.* Chicago: Quadrangle Books.

Corbet, P.S. 1999. *Dragonflies: Behavior and ecology of Odonata.* Ithaca, N.Y.: Cornell University Press. An exceptionally well done and encyclopedic reference to all aspects of dragonfly life.

Dunkle, S.W. 2000. *Dragonflies through binoculars: A field guide to dragonflies of North America.* New York: Oxford University Press. Photographs and identification tips for all North American anisopterans.

Garrison, R.W. 1984. *Revision of the genus* Enallagma *of the United States west of the Rocky Mountains and identification of certain larvae by discriminant analysis.* University Of California Publication, Entomology, no. 105. Berkeley: University of California Press. Detailed information on distribution and identification of all western bluets, including larvae.

Garrison, R.W. 1994. A synopsis of the genus *Argia* of the United States, with keys and descriptions of new species *Argia sabina, A. leonorae,* and *A. pima* (Odonata: Coenagrionidae). *Trans.*

Am. Entomol. Soc. 120:287–368. Detailed information on all western species of dancers, including many illustrations useful for identification of similar species.

Kennedy, C.H. 1915. Notes on the life history and ecology of the dragonflies (Odonata) of Washington and Oregon. *Proc. U. S. Natl. Mus.* 49:295–345. Useful information on natural history of species that occur in California, too.

Kennedy, C.H. 1917. Notes on the life history and ecology of the dragonflies (Odonata) of central California and Nevada. *Proc. U.S. Natl. Mus.* 52:483–635. A classic, containing a wealth of information on the natural history of the Odonata of California.

Needham, J.G., M.J. Westfall Jr., and M.L. May. 2000. *Dragonflies of North America.* Gainsville, Fla.: Scientific Publishers. The definitive, up-to-date handbook on North American anisopterans.

Nikula, B., J. Sones, D. Stokes, and L. Stokes. 2002. *Stokes beginner's guide to dragonflies and damselflies.* Boston: Little, Brown.

Paulson, D. 1999. *Dragonflies of Washington.* Seattle: Seattle Audubon Society. Photographs and identification tips for many species shared with California.

Paulson, D.R., and S.W. Dunkle. 1999. *A checklist of North American Odonata, including English name, etymology, type locality, and distribution.* Occasional Paper no. 56. Tacoma, Wash.: Slater Museum of Natural History, University of Puget Sound. Basic information on the meanings of the English and scientific names of all North American species.

Paulson, D.R., and R.W. Garrison. 1977. A list and new distributional records of Pacific Coast Odonata. Pan-Pac. Entomol. 53:147–160.

Powell, J.A., and C.L. Hogue. 1979. *California insects.* Berkeley: University of California Press. A basic introduction to the insect fauna of California, including information on making an insect collection.

Pritchard, D.R., and R.F. Smith. 1956. Odonata. In *Aquatic insects of California,* edited by R.L. Usinger. Berkeley: University of California Press. Out of date and out of print, but with useful keys, including larval keys.

Walker, E.M. 1953. *The Odonata of Canada and Alaska.* Vol. 1. Toronto: University of Toronto Press.

Walker, E.M. 1958. *The Odonata of Canada and Alaska.* Vol. 2. Toronto: University of Toronto Press.

Walker, E.M., and P.S. Corbet. 1975. *The Odonata of Canada and Alaska.* Vol. 3. Toronto: University of Toronto Press.

Westfall, M.J. Jr., and M.L. May. 1996. *Damselflies of North America.* Gainsville, Fla.: Scientific Publishers. The definitive handbook to the North American zygopterans. Includes keys, some photographs, and information on collecting.

INDEX

Page references in **boldface** refer to the main discussion of the species.

ABOUT THE AUTHOR

Timothy D. Manolis is an artist, illustrator, and biological consultant whose writings about birds have appeared in many journals. He received a Ph.D. in biology from the University of Colorado, and is a former editor and art director of *Mainstream* magazine. He resides in Sacramento, California.

Series Design:	Barbara Jellow
Design Enhancements:	Beth Hansen
Design Development:	Jane Tenenbaum
Illustrator:	Tim Manolis
Cartographer:	Bill Nelson
Composition:	Impressions Book and Journal Services, Inc.
Text:	9.5/12 Minion
Display:	ITC Franklin Gothic Book and Demi
Printer and Binder:	Everbest Printing Company

CALIFORNIA NATURAL HISTORY GUIDES

"It's always good to read a new California Natural History Guide; these little books are small enough to fit into a pocket, inexpensive, and authoritative." —*Sunset*

"A series of excellent pocket books, carefully researched, clearly written, and handsomely illustrated." —*Los Angeles Times*

The California Natural History Guide series is the state's most authoritative resource for helping outdoor enthusiasts and professionals appreciate the wonderful natural resources of their state. If you would like to receive more information about the series or other books on California natural history, please fill in this card and return it to the University of California Press or register online at www.californianaturalhistory.com.

Name _____

Address _____

City/State/Zip _____

Email _____

Which book did this card come from? _____

_____

Where did you buy this book? _____

_____

What is your profession? _____

UNIVERSITY OF CALIFORNIA PRESS
www.ucpress.edu

Return to:
University of California Press
Attn: Natural History Editor
2120 Berkeley Way
Berkeley, California 94720